Astrobiology

The Search for Life in the Universe

Arnold Hanslmeier
Institute of Physics
Univ. Graz, Austria

Contents

FOREWORD

Astrobiology is a relatively new science focusing on the study of the Origin of Life, evolution, biodiversity, and the future of life in the universe. This domain gathers scientists from various fields (microbiology, ecology, geology, chemistry, biochemistry, astronomy, astrophysics and paleontology). Astrobiology tries to answer the questions of what is life? Are we alone in the Universe? In addition, this science devotes deep analysis to the extraterrestrial worlds and where does life possibly exists. This volume deals with several aspects of the astrobiological questions about "life" on our planet and elsewhere in the Universe. The activities toward searching for and finding life in the Universe is a fascinating topic, albeit not an easy one. Life exists almost all over our Earth, in very diversified ranges of habitats, from "normal" fields to extreme environments. The subdivisions in the ten main chapters of this book allow the subjects to be examined in depth., from the first chapter with "what is life" to the last, with "Search of Life". The author has penetrated deeply into the range of Astrobiology topics, so the reader gains a thorough grounding in each of them-the author has composed a veritable encyclopedia for this field.. The Earth position embedded in the solar system, its distance from the sun and its surrounding with the encircling atmosphere is similar to the fairy tale of Goldilocks case. These factors (position in space and the atmosphere of Earth) protect life by providing ambient temperature and by avoiding the penetration of energy particles, energy radiation, UV radiation (via the ozone screen), and by keeping the temperature in a normal level, and irradiating with the right wavelengths of visible light and save our planet from other negative factors. From the solar system's eight planets and several satellites (or moons), some of might be candidates for habitable places. The prerequisites for signs of life are the presence of liquid water, an energy source, and complex carbon-based structure. There is currently indeed an active search for habitable zones among the Solar System and extrasolar planets. The assumption is that in the huge cosmos, there must be civilizations similar to what we have on Earth. Furthermore, among the celestial bodies of the solar system are promising chances to detect Life, mainly in the subsurface of places such as Mars, Europa, Titan and other extraterrestrial places. Mars itself was once warm and wet with lakes, rivers, canyons and gullies where water used to run (based on the photos of the flyby space crafts and rovers on the surface). NASA (2006) reported that water has been flowing at the gullies on crater walls in the southern Martian pole. In Vostok station in Antarctica, there exists an ocean four km under the ice layer. A similar but much larger subsurface ocean exists in Europa (moon of Jupiter). This Europa Sea is found at a depth of a few kilometers under the frozen icy cover and is estimated to be a 100-kilometer-thick layer of water. Titan, the large satellite of Saturn, has dense atmosphere (pressure of 1.5 Bar) of nitrogen and methane and lakes of liquid ethane and methane; the surface temperature is - 179C. Our information about Venus does not let us consider it as habitable body. In the past, however, Libby (1968) built a model where ice caps are at the poles and in a certain region, this planet may carry life (Sagan, 1967:

Seckbach and Libby, 1970, 1971). The extremophiles (organisms thriving in very severe environments) might be good analogues or models for extraterrestrial life. Among these extremophiles are hyperthermophiles (tolerating very high temperature levels, up to 160C in deep sea hydrothermal vents), cryophyles/psychrophiles (microbial growing in very cold temperatures, of ?20), acidophiles (microbes living in pH 0 to 4), alkaliphiles (thriving at pH 8-11), barophiles (organisms under high pressures), xerophiles (adapted to grow under limited supply of water), halophiles (microorganisms that need high salt concentration to grow) (see Seckbach, 2007; Seckbach, et al. 2013). These terrestrial extremophiles might be in active metabolism or at rest in a dormant stage even for million years in isolated subsurface, mines, salt crystals, under glaciers or at the bottom of the oceans. An active search for celestial life would be for life types that are characteristic of "life as we know it" and the passive search, for "life as we do not know it." Using huge antenna dishes the SETI group is searching for intelligent civilizations with radio message broadcast into the huge galaxy neighborhood. Since the target distance toward exosolar systems is many light years away from us, the hope is to receive a reply during the next years. The reading audience of this book should include graduate students and scientists (in the fields of astronomy, astrophysics, astrobiology, geology, biology) as well as interested "open-minded" readers. Each chapter provides many references for further elucidation presented of the topics offered. The astrophysicist author of this book assembled collections of updated scientific data which complement his publications of other books in related areas. It is hoped that the readers will gainfully extract astrobiological information from this work and also enjoy the colorful figures and illustrations accompanying the text.

Joseph Seckbach Hebrew University of Jerusalem, Jerusalem, IL.

Preface

The search for life in the universe is one of the most difficult but fascinating topics in science and perhaps one of the most challenging questions of humans. Are we alone in the huge universe? From theoretical arguments presented in the last chapters these seems to be quite unlikely. But before elucidating on such a complex problem we have to answer the much simpler question of what is life? The answer is given in the first chapter but it is not an easy one. Life is extremely complex. We are familiar only with the life on Earth. This life has certain properties, being based on the presence of liquid water and on complex carbon based structures. This is reviewed in the first two chapters, however, could life exist under completely different conditions not based either on water or on carbon? There are strong reasons to believe that this is unlikely but has not yet been proven.

In the first three chapters of this book we will first give some insight into how life functions on Earth, what are the basic properties and how life may have originated on Earth. Our Earth acts also as a protector for life. Its atmosphere prevents hazardous short wavelength radiation (X rays or UV) to reach the surface of Earth and the magnetic fields protects the Earth from energetic solar particles and other high energy charged particles from the cosmos. Moving further the chapter gives of the Solar System and the possibility for finding life on any of these bodies.

One of the great advances during the past two decades in astrophysics has been the discovery of planets outside the Solar System. Methods for finding such objects as well as their results, the central stars of these exoplanetary systems are discussed. Together with the intent to answer the question whether these objects might be habitable. This culminates in the discussion on habitability. What makes an object habitable? Several definitions of habitable zones are presented followed by a short review of the process of element synthesis in stellar cores and also the Big Bang theory according to which the two most abundant elements in the universe, hydrogen and helium, were formed during the first minutes of the evolution of the universe. The last chapter is about active and passive search for life with theoretical arguments given to estimate how many civilizations there might exist in our Galaxy. Passive search also includes the broadcasting of radio messages to possible civilizations and several attempts to focus closely on different stars.

The main intention of the book is to assist the reader to perform his/her own estimations and calculations applying several examples given throughout the text. At the end of each chapter a section called "activities" is given answering more complex questions.

The book is written for scientists, students, interested readers with some background on science in general. Each chapter goes from a general point to details including more than 100 literature citations. These allow a deeper penetration into the subjects presented.

I want to thank my colleagues for the cooperation, especially Dr. H. Lammer and his group, most of them also being my students. I also thank my girl friend Anita for her continued support interest and patience. Wikipedia and the anonymous colleagues that provided are wonderful resources which are acknowledged in general as well as the NASA ADS system where a quick overview on Literature can be found.

Finally I want to thank my publisher and proofreaders for their extremely helpful comments and wish the reader many exciting and fascinating hours with this book. The cover image was taken from NASA/JPL.

The author confirms that this eBook content has no conflicts of interest.

Arnold Hanslmeier,
Institute of Physics, Univ. Graz, Austria

CHAPTER 1

What is life?

Abstrac: This chapter opens with the very difficult assessment of life. A careful definition is necessary to know how to search for the existence of life in the universe, especially on other planets. Until the end of the twentieth century it was hardly proven from observations that planets outside our Solar System exist. However, with the technological advancement we know more than 1000 extrasolar planets, with the number increasing steeply because of the new satellite missions (see chapter on extrasolar planets). The main problem however is, that till now we only know about the existence of life on Earth. Could life evolve elsewhere under completely different conditions than on Earth? Based on our knowledge on terrestrial life, we will also give some basic information on how living cells function. We give an overview on how life works on Earth by discussing photosynthesis, respiration, organic molecules, molecules that are essential for life. The complex organic compounds such as carbohydrates, proteins, amino acids are discussed and how they work together in a cell. Cell division is also reviewed and it is demonstrated why this mechanism has evolved from a simple fusion process to such a complex process like meiosis. The main message of this chapter is that life is complex, based on complex molecules, reproduces itself and needs energy.

Keywords: Life; definition of life; prokaryotes; eukaryotes; cells; organic molecules; photosynthesis; respiration; cell division; mitosis; meiosis.

Preparing Activities

These activities are aimed to discuss the definition of life and how to start a hypothetical search for life. For readers not acquainted with the term extrasolar planet, the definition goes as: Our Solar System contains the Sun as host star and eight large planets including Earth. Extrasolar planets or exoplanets are planets those revolve around a host star different from the Sun.

1. Try to find your own definition of life.

2. What is an extrasolar planet?

3. Is there a limit for the size of a living organism?

4. Imagine you want to search for life on the Moon. Which experiments would you suggest?

1.1   Property and Definition of Life

In the later chapters we will define a habitable zone around a star or planet. Such a definition requires a general concept of what is life.

1.1.1   Life - how it is defined

Can life be considered as a self-sustaining chemical system turning environmental resources into its own building blocks? Living beings can be regarded as thermodynamic systems. Like time, entropy runs in one direction. In physics, the second law of thermodynamics states that the entropy of a closed system always increases or stays constant. It is a measure for the tendency toward equilibrium/average/homogenization/dissipation in nature:

- hotter, more dynamic areas of a system lose heat/energy while cooler areas (*e.g.*, space) get warmer / gain energy; a cup of hot coffee gets colder warming the air surrounding it; it will never occur that the surrounding air cools and the coffee becomes warmer automatically.

- molecules of a solvent or gas tend to evenly distribute;

Such entropy reduction occurs locally in an organism however entropy reduction requires external energy or chemical substances. On a more global scale, entropy increases. Such considerations were made by different physicists like Schrdinger or Wigner.

At a first sight it seems to be trivial to describe life but the examples below show that life cannot be defined by some simple properties, thereby making the answer quite complex.

Let us give some statements and make comments to them.

`Living organisms grow.`

This is a true statement but not sufficient enough considering the growing crystals as nobody would consider them as living, however they grow.

`Movement of living things.`

Non living objects even move under the influence of external forces like a pellet of dry ice (frozen carbon dioxide) dropped into water. The pellet will move randomly across the surface.

Therefore this necessitates a worthy definition. For terrestrial life, cells are the building blocks. These cells are mainly composed of the cytoplasm bound by a very thin membrane. Moreover, living cells contain genetic material. This controls their evolution and activities. In many cells the genetic material, DNA, is found in the nucleus, a spherical structure suspended in the cytoplasm. In simpler forms of life such as bacteria, the DNA is distributed in the cytoplasm. Fig. 1.1 shows typical plant cells with some organelles. The cellular structure of life is one aspect, but there are further criteria to be taken into consideration.

- The need for *energy*: living thinks cannot carry on activities without an outside source of energy and materials. Trees for example use carbon dioxide, water, and solar energy to produce their own food. Animals contrarily acquire materials and energy when they eat food.

- *Growth*: new cells are produced. Growth means an increase in the size and often the number of cells.

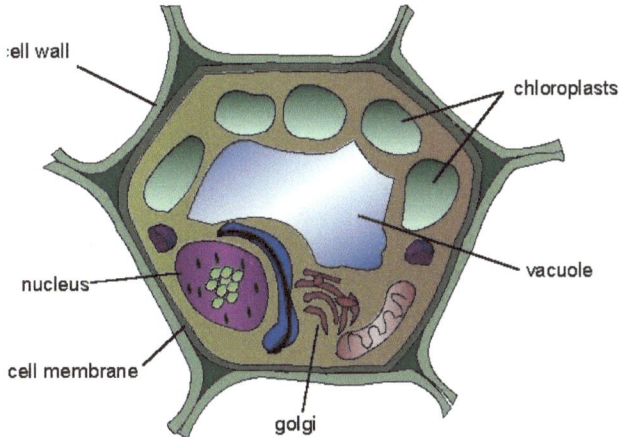

Figure 1.1: A typical plant cell. The Golgi apparatus is important for cell secretion. Chloroplasts are found in plant cells and other eukaryotic organisms. They conduct photosynthesis. The vacuole has different functions such as isolating materials that might be harmful or a threat to the cell containing waste products or water in plant cells.

- *Reproduction*: when organisms reproduce, the offspring resemble the parents. Genes are in the form of DNA molecules allowing cells and organisms to reproduce. The DNA contains hereditary information. Before reproduction occurs, DNA is replicated (see Fig. 1.2), introducing exact copies of genes. Unicellular organisms reproduce asexually by dividing. The new cells have the same genes and structure as the single parent. Multicellular organisms usually reproduce sexually. Each parent, male and female, contributes about one-half of the total number of genes to the offspring. The main progress in the evolution of this is the fact that the offspring does not resemble either parent exactly, therefore mutations become possible.

- *Response* to stimuli: this is a major characteristic of all living things. Plant responses to stimuli are generally much slower than those of animals. Some leaves of plants track the passage of the sun during the day. When a plant is placed near a window, its stem bends to face the sun. This shows that behavior is largely directed toward minimizing injury, acquiring food, and reproducing.

- *Homeostasis*: this Greek word means "staying the same". The internal environment of living things stays relatively constant. In the human body temperature keeps on fluctuating slightly during the day. All organ systems contribute to homeostasis such as the digestive system provides nutrient molecules, the cardiovascular system transports them around the body and the urinary system gives off metabolic wastes.

- *Metabolism*: collective product of all the biochemical reactions taking place within an organism. Metabolism includes photosynthesis, respiration, digestion and assimilation. New cytoplasm is produced, damage is repaired and normal cells are maintained.

- *Movement*: also plants can move. Leaves of sensitive plants (*e.g.* Mimosa pudica) fold within a few seconds after being disturbed or subjected to sudden environmental changes.

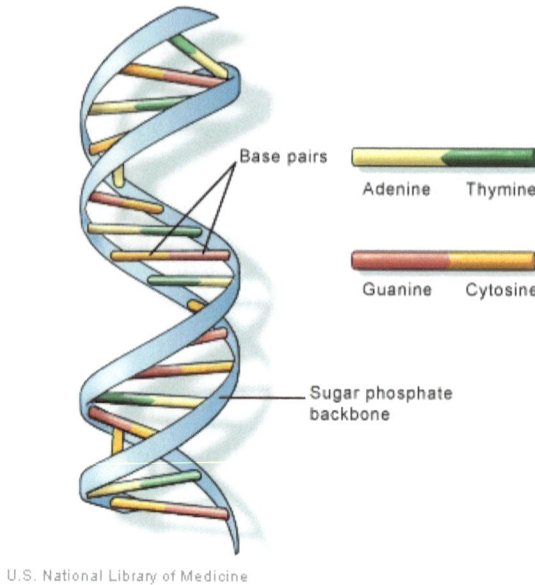

Figure 1.2: The DNA is a double helix formed by base pairs attached to a sugar-phosphate backbone. Adapted from: US Nat. Libr. of Medicine.

- *Complexity* of organization: cells are composed of large numbers of molecules (generally more than one trillion in a typical cell). The molecules are organized into compartments, membranes and other structures in the cell. Bacteria are considered to have the simplest cells known, yet this cell contains at least 600 different kinds of proteins and 100 other substances. Other organisms are more complex.

- *Adaption* to the environment: living organisms respond to their environment, to air, light, water, soil *etc*. Natural selection leads to adaption of organisms to their environment. Whenever a new variation arises that allows certain members of the species to capture more resources, these members tend to survive and have more offspring than the other, being unchanged members. This drives evolution. Evolution describes both the unity and the diversity of life. All organisms share the same characteristics of life because their ancestry can be traced back to the first cell or cells. Organisms are diverse because they are adapted to different ways of life and environment. Today many species are threatened with extinction because they are not able to adapt fast enough to the changing environment.

The above made statements about properties of life are reflected in the definition of life that can be found on NASA's astrobiology website: Living things tend to be complex and highly organized. They have the ability to take in energy from the environment and transform it for growth and reproduction. Organisms tend toward homeostasis: an equilibrium of parameters that define their internal environment. Living creatures respond, and their stimulation fosters a reaction-like motion, recoil, and in advanced forms, learning. Life is reproductive, as some kind of copying is needed for evolution to take hold through a population's mutation and natural selection. To grow and develop, living creatures need foremost

Figure 1.3: Exterior view of Biosphere 2, an attempt to create a closed, artificial ecosystem on an area of more than 12,000 m^2 in Oracle, Arizona. It also explored the possible use of closed biospheres in space colonization. Today, it is a tourist attraction.

to be consumers, since growth includes changing biomass, creating new individuals, and the shedding of waste.

Note that the definition of life is a complex one and cannot be given in just one or two sentences.

1.1.2   The biosphere

The biosphere is the zone of air, land, and water at the surface of the Earth where living organisms are found. Individual organisms belong to a population. All members within a specific area belong to a population. The populations of a community interact among themselves and with the physical environment (water, atmosphere, soil,...), forming an ecosystem.

Manned long time space missions require an autonomous supply for food. The spaceship should ideally function as a biosphere where it runs a complete ecosystem. Several tests of such systems were conducted (see for example Fig. 1.3).

Let us consider a grass land inhabited by populations of rabbits, mice, snakes, and various types of grasses. These populations interact with the atmosphere, interchange gases, and give off heat. They absorb water and give off water. The populations interact with each other by forming food chains with one population feeding the other. Mice feed on plant products, snakes feed on mice *etc.*

Ecosystems are characterized by chemical cycling and energy flow. Photsynthesizers take in solar energy and inorganic nutrients to produce food (organic nutrients), gradually dissipating energy, which is known to flow rather than to cycle. Ecosystems could not stay without a supply of solar energy. On Earth there is a complete cycle between producers of food and consumers, as is illustrated in Fig. 1.4.

Climate determines where different ecosystems are found examples being deserts, tropical rain forest or coral reefs.

Biodiversity encompasses the total number of species, the variability of their genes, and the ecosystems in which they live. The present biodiversity of our planet has been estimated to be as high as 15 million species, and so far, about 2 million have been identified and named. Extinction is the death of a species or larger taxonomic group. Presently, we are losing as many as 400 species per day due to anthropogenic effects. During our planet's history, five

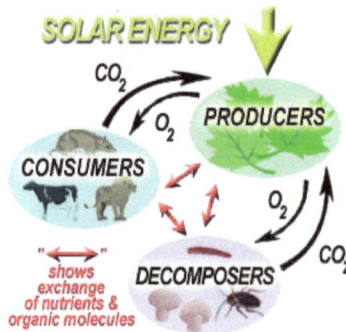

Figure 1.4: An example of a cycle, how producers (plants) produce oxygen that consumers (animals) need. The consumers give off CO_2 that plants need. From:
http://www.bigelow.org/bacteria/

major mass extinctions have occurred. The dinosaurs became extinct during the last mass
extinction, 65 million years ago.

The study of biospheres and ecosystems helps us to understand how complex systems
behave and work together.

1.1.3   Levels of Biological Organization

A cell is the smallest structural and functional unit of an organism. A nerve cell has a
structure suitable to conducting a nerve impulse. A tissue cell is a group of similar cells
that perform a particular function. Several tissues join together to form an organ. Organs
work together to form an organ system. In the human body, the brain sends messages to
the spinal cord, which, in turn, sends them to body parts by way of spinal nerves. Complex
organisms such as humans are a collection of organ systems. The human body contains a
digestive system, a cardiovascular system, a nervous system and others.

At the lowest level of organization, atoms join to form the molecules that are found in
cells. The cells are the smallest unity of life and differ in shape, size, and function. Tissues
are group of cells, organs are composed of different tissues and make up the organ system.

The increasing complexity of life from atoms to cells to complex organisms is illustrated
in Fig. 1.5.

1.1.4   Cells

The cell was discovered by Robert Hooke in 1665. He saw cork cells through his microscope.
Being the basic unit of all known living organisms the cell is often called the building block
of life. Organisms can be classified as

1. unicellular: consisting of a single cell; including most bacteria

2. multicellular: including plants, animals.

Humans contain 100 trillion cells. A typical cell size is 10 μm and a typical cell mass is 1
nanogram. The largest known cells are unfertilised ostrich egg cells which weigh 3.3 pounds.

Non living   living

Organ structures

Organs

Tissues

Atoms

Unicellular   multicellular

<mark>Cells</mark>

Molecules

Complexity of life increases

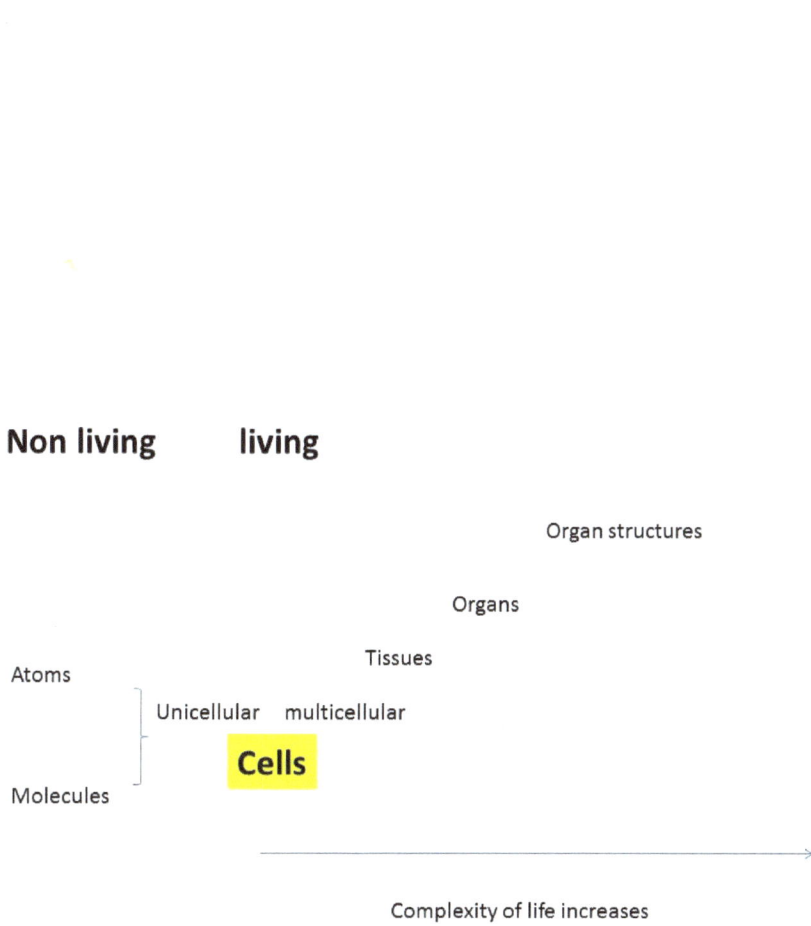

Figure 1.5: Increasing complexity of life. The complexity of life on Earth increased with time but not linearly.

Table 1.1: Comparison of prokaryotic and eukaryotic cells.

	Prokaryotes	Eukaryotes
Typical organisms	bacteria, archaea	protists, fungi, plants, animals
Typical size	$1 - 10\,\mu m$	$10 - 100\,\mu m$
Type of nucleus	no real nucleus	real nucleus with membrane
Cytoplasmatic structure	very few	highly structured
Mitochondria	none	one to several thousand
Chloroplasts	none	in algae and plants
Cell division	fission	Mitosis, Meiosis

These criteria may facilitate us to start the search for life. What type of experiments would you suggest to search for life signatures on Mars?

There are two types of cells: eukaryotic and prokaryotic(Fig. 1.6). The prokaryotic cells are usually independent while eucaryotic cells are often found in multicellular organisms. The main differences between these two types are listed in Table 1.1.

The mitochondria are self-replicating organelles that occur in the cytoplasms of all eucaryotic cells. They play a critical role in generating energy. They use oxygen to release energy stored in cellular nutrients (usually glucose) to generate ATP.

Describe the importance of energy for living organisms. How do different types of living forms (plants, animals) get their energy?

1.1.5 Cell Division

Cell division has to be seen in the context of evolution and reproduction of life. Errors occurring during cell division lead to mutations and are a main driver for evolution.

A human being's body experiences about 10,000 trillion cell divisions in a lifetime. The primary reason for cell division is the maintenance of the original cell's genome. Before division can occur, the genomic information stored in the chromosomes must be replicated, and the duplicated genome separated cleanly between cells.

The simplest way of cell division is *binary fission* which is typical for organisms having a single chromosome such as bacteria. At the start of the binary fission process, the DNA molecule of the cell's chromosome is replicated, producing two copies of the chromosome. A key aspect of bacterial cell reproduction is making sure that each daughter cell gets a copy of the chromosome.

Mitosis is a process of cell division resulting in the production of two daughter cells from a single parent cell. The daughter cells are identical to one another and to the original parent cell. The process is based on four phases:

- Prophase: The chromatin diffuses in interphase, condensing into chromosomes. Each chromosome duplicates and now consists of two sister chromatids. At the end of prophase, the nuclear envelope breaks down into vesicles.

- Metaphase: The chromosomes align at the equatorial plate.

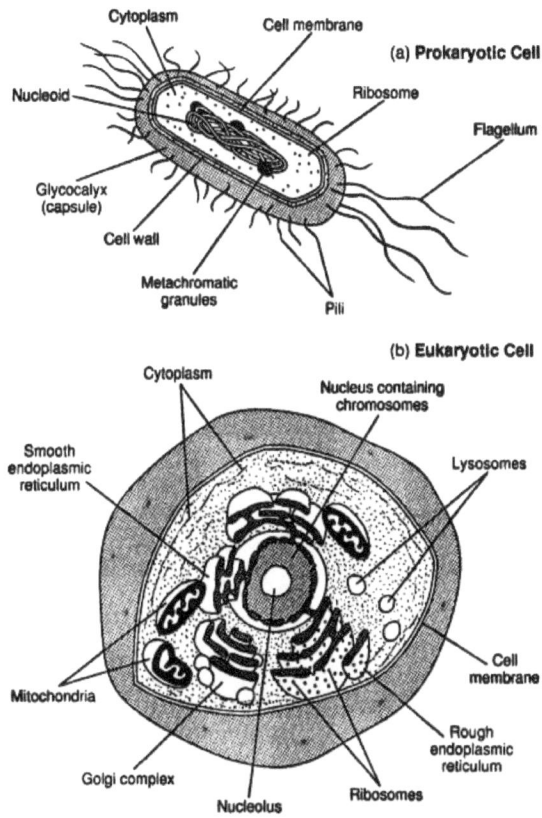

Figure 1.6: Prokaryotic and eukaryotic cell. From: CliffsNotes

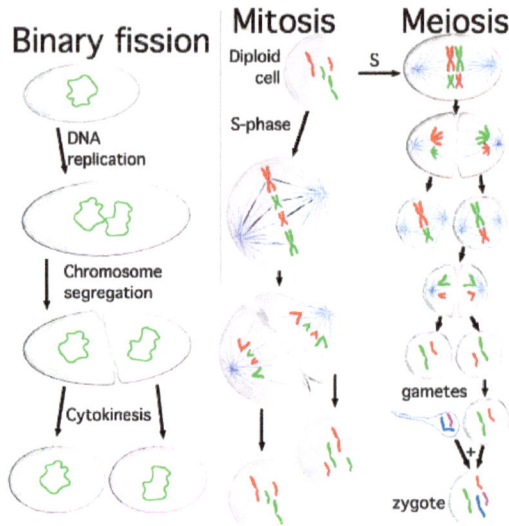

Figure 1.7: Three types of cell division: fission, mitosis, meiosis.

- Anaphase: The centromeres divide and sister chromatids separate and move toward the corresponding poles.

- Telophase: Daughter chromosomes arrive at the poles and the microtubules disappear. The cytoplasm divides, and two daughter cells are produced (phase: Cytokinesis).

The process, during which the germ cells are generated is called *meiosis*. Diploid cells are the cells with a double set of chromosomes. If two diploid cells would fuse, chromosome doubling would occur. Accordingly, meiosis produces haploid germ cells, with maternal and paternal germ cell fusing at fertilization and thus generating a diploid fusion product, the zygote. The main stages are:

- gametes: production of haploid cells,

- fertilization, fusion of sperm + egg

- zygotes: diploid, contain information from both parent cells

- zygotes: through numerous mitoses, multicellular organisms evolve

In Fig. 1.7 the three types of cell division are summarized. It is evident that meiosis is the most complicated procedure however it enables the evolution of modified species more easily than mitosis or cell fission. Therefore, evolution depends on the complexity of cell division mechanisms. In multicellular organisms cell division eventually halts. In humans, this occurs on average, after 52 divisions, known as the Hayflick limit[1]. The cell is then referred to as senescent. Some cells become senescent after fewer replication cycles as a result of DNA double strand breaks, toxins, *etc.*

Organismal senescence is the aging of whole organisms. In general, aging is characterized by the declining ability to respond to stress, increased homeostatic imbalance, and increased risk of aging-associated diseases.

---

[1]L. Hayflick, 1961

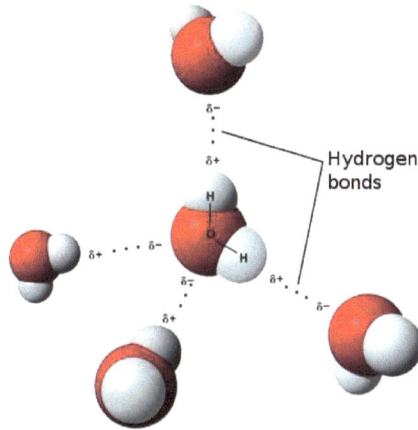

Figure 1.8: Water molecules consist of oxygen (red) and two hydrogen atoms (small, white). The hydrogen bonds act between the slightly positively charged hydrogen and negatively charged oxygen atoms.

Discuss why nature has evolved different strategies for cell division. What is aging? How can aging of cells be affected?

1.2   Molecules Important for Life

Inorganic molecules constitute the nonliving matter. Several inorganic molecules such as salts or water play important roles in living things. The molecules of life from Earth are organic molecules and they always contain carbon and hydrogen.

1.2.1   Some Important Inorganic Molecules for Life

Water is the most abundant molecule in living organisms making up 60-70 % of the total body weight. Water consists of two hydrogen (H) and one oxygen (O) atoms. The electrons spend more time circling the larger oxygen atom than the smaller H atom. Therefore they impart a slight negative charge to the oxygen and a slight positive charge to the hydrogen atoms. The water molecule has a positive and a negative end and becomes a polar molecule. Between the slightly negative oxygen atom of a particular water molecule and the slightly positive charged hydrogen atom of a neighboring molecule there exists a small attraction which is called a hydrogen bond (see Fig. 1.8).

Because of their polarity and hydrogen bonding, water molecules are cohesive, enabling them to cling together. This gives water the characteristics beneficial to life. Water is liquid at room temperature and above until 100^0 C [2]. It remains liquid in our bodies. Water is a universal solvent for polar molecules and thereby facilitates chemical reactions both outside and inside our bodies. Ions and molecules that interact with water are said to be hydrophilic. Non- ionized and non polar molecules that do not interact with water are said to be hydrophobic. Water molecules are cohesive, and therefore water based solutions fill

---

[2]This depends on the external pressure; on Mars, temperatures slightly above 0^0 are reached but because of the low Martian atmospheric pressure liquid water would immediately evaporate

Table 1.2: Significant ions in the body.

Name	Symbol	Significance
Sodium	Na^+	body fluids, muscle contraction, nerve conduction
Chloride	Cl^-	body fluids
Potassium	K^+	inside cells; muscle contraction, nerve conduction
Phosphate	PO_4^{3-}	bones, teeth, ATP
Calcium	Ca^{2+}	bones, teeth, nerve conduction, muscle contraction
Bicarbonate	HCO_3^-	acid-base balance
Hydrogen	H^+	acid-base balance
Hydroxide	OH^-	acid-base balance

vessels, such as blood vessels. Blood in the human body consists of 92 % of water. Blood transports oxygen and nutrients to the cells and removes wastes such as carbon dioxide from the cells. One further property of water is that the temperature of liquid water rises and falls slowly. Therefore, water protects the organisms from rapid temperature changes and helps to maintain internal temperature of an organism. Water is an excellent temperature buffer. Great bodies of water, such as oceans, maintain a relatively constant temperature. To raise the temperature of 1 g of water by 1 ^{0}C, one calorie is needed (1cal =4,184 J). To convert one gram of the hottest water to steam requires 540 cal. This also helps to moderate the Earth's temperature. Frozen water is less dense than liquid water. As water cools, the molecules come close together becoming densest at 4^0 C. Bodies of water like lakes or rivers always freeze from the top to the bottom. When a body freezes at the surface, the ice acts as an insulator to prevent the water below from getting frozen. Without these properties the aquatic organisms would not survive the winter. Organisms are prone to adapt themselves to changing conditions. Some produce antifreeze chemicals that maintain their body fluids liquid while other organisms dehydrate and are able to survive.

Most conceivable metabolism reactions are kinetically allowed in the aqueous phase making water essential for life.

What makes water a unique molecule for life?

A list of significant ions in the body is given in Table 1.2.

Further important molecules for life on Earth are CO_2, O_2, HCN, and others.

CO_2 is one product of respiration where O_2 is inhaled and this oxygen is needed for the production of energy.

- In physiology, respiration [3] is defined as the transport of O_2 from the outside air to the cells within tissues, and the transport of CO_2 in the opposite direction.

- Biochemical definition of respiration, refers to cellular respiration: the metabolic process by which an organism obtains energy by reacting oxygen with glucose to give water, carbon dioxide and ATP (energy).

Hydrogen cyanide, HCN, being a colorless, extremely poisonous liquid that boils slightly above room temperature at 26^0C has been discussed as a precursor to amino acids and nucleic acids. This molecule has also been detected in the interstellar medium.

---

[3]often confused with breathing

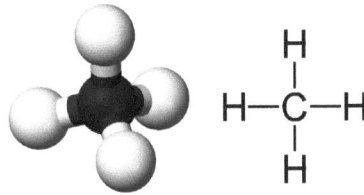

Figure 1.9: Methane.

Table 1.3: Monomers and polymers in biology

Polymer	Monomer
carbohydrate (*e.g.*, starch)	monosaccharide
protein	amino acid
nulceic acid	nucleotide

1.2.2 Organic Molecules

Organic molecules always contain hydrogen and carbon. A carbon atom has four electrons in the outer shell which can be filled up with 8 electrons. In order to achieve eight electrons in the outer shell, a carbon atom shares electrons covalently with up to four atoms. A typical example is methane, CH_4 (Fig. 1.9).

If you want to make long distance interstellar travel, is it possible to freeze the body until arrival to target star?

A carbon atom can share electrons with another carbon atom and a long hydrocarbon chain can result. This chain can turn back on itself to form a ring compound. Functional groups can be attached to carbon chains such as the carboxyl group −COOH.

Living matter consists of macromolecules. There are many molecules joined together. A monomer is a simple organic molecule that can exist (i) individually or (ii) link with other monomers to form a polymer. For biology, the following monomers and polymers are important (see Table 1.3).

- carbohydrates

- proteins

- lipids

- nucleic acids

The nucleic acid DNA makes up our genes. These control our cells and the structure of an organism.

How are monomers joined together to build polymers? This occurs by a dehydration action: OH (hydroxyl group) and H are removed. These correspond to a water molecule. In cells, synthesis occurs when subunits bond following a dehydration reaction (removal of H_2O). Degradation occurs when the subunits in a macromolecule separate after the addition of H_2O.

glucose   **fructose**

sucrose

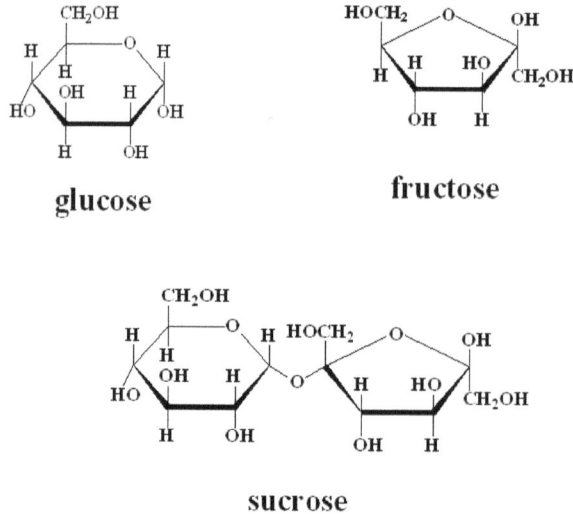

Figure 1.10: Sucrose, one of the sugars found in soybeans, is easily digestible. The sucrose molecule is made up of two simpler sugars called glucose and fructose. One glucose molecule bonded to one fructose molecule makes one sucrose molecule.

1.2.3   Carbohydrates

Carbohydrates mainly store energy. They are characterized by the presence of the atomic grouping $H - C - OH$. The ratio of hydrogen atoms to oxygen atoms is approximately 2:1, which is the same as the ratio in water, hence the name carbohydrates. If the number of carbon atoms is low (3...7), the carbohydrate is a simple sugar (e.g. Fig. 1.10) or a monosaccharide. Pentose means a 5-carbon sugar, while hexose means a 6-carbon sugar. Similarly, glucose is a hexose, blood sugar. Our bodies use glucose as an immediate source of energy. Other common hexoses are fructose (in fruits) and galactose (in milk). These three hexoses (glucose, fructose and galactose) occur in rings with the molecular formula being $C_6H_{12}O_6$.

What are carbohydrates? What is glucose?

Let us discuss the synthesis of maltose.

- dehydration reaction:

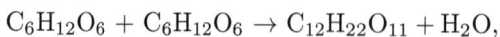

  $C_6H_{12}O_6 + C_6H_{12}O_6 \rightarrow C_{12}H_{22}O_{11} + H_2O,$

  which means: glucose+ glucose $\rightarrow$ maltose, or monosaccharide + monosaccharide $\rightarrow$ disaccharide.

- hydrolysis reaction:

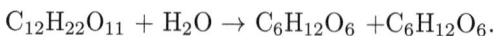

  $C_{12}H_{22}O_{11} + H_2O \rightarrow C_6H_{12}O_6 + C_6H_{12}O_6.$

A disaccharide contains two monosaccharides joined during a dehydration reaction. In our body the hydrolytic digestive juices can break the bond in the disaccharide and two glucose molecules result.

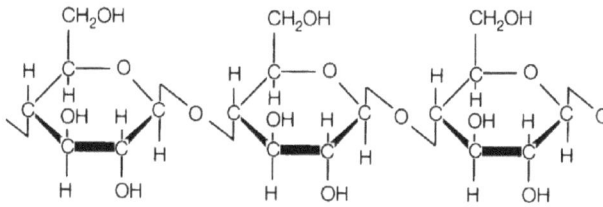

Figure 1.11: Cellulose, a complex organic molecule.

When glucose and fructose join, the disaccharide sucrose is formed. This is the usual sugar from plants (sugar beets or sugarcane). Lactose is combined glucose and galactose.

The carbohydrates discussed so far are simple carbohydrates. Long polymers such as starch, glycogen, and cellulose are polysaccharides that contain many glucose subunits. Starch and glycogen are polymers. Starch has polymers as long as 4000 glucose units. After eating starchy foods (potatoes, bread, cake) glucose enters the bloodstream and the liver stores glucose as glycogen. In between eating, the liver releases glucose so that the blood glucose concentration is always about 0.1 %

The polysaccharide cellulose (Fig. 1.11) is found in plant cell walls. The glucose units are joined by a slightly different type of linkage than that found in starch or glycogen, which retards the digestion of foods containing this type of linkage. Cellulose, however, passes through our digestive tract as fiber.

What is cellulose? What is a disaccharide? What are polysaccharides?

1.2.4   Proteins

Proteins perform many functions. Keratin basically makes up hair and nails, while collagen gives support to ligaments tendons and skin. Hormones are messengers that influence cellular metabolism, as many hormones are proteins. Actin and myosin account for the movement of cells and muscle contraction. On the other hand, Hemoglobin being a complex protein transports oxygen in the blood. Antibodies , however, combine with foreign substances preventing them from destroying cells. In the plasma membrane, proteins form channels that allow substances to enter and exit cells. Enzymes are responsible to speed chemical reactions. A reaction that normally takes several hours without an enzyme takes only a fraction of a second, due to the secretion of enzymes.

Proteins are polymers with amino acid (Fig. 1.12) and monomers. An amino acid has a central carbon atom bounded to a hydrogen atom and three groups:

- Amino group $-NH_2$

- acidic group $-COOH$

- R group; R stands for remainder.

A condensation synthesis reaction between two amino acids results in a dipeptide and a molecule of water. A polypeptide is a single chain of amino acids.

The structure of a protein has several levels of organization.

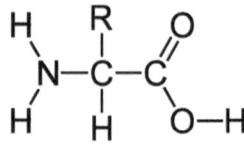

Figure 1.12: Alpha-amino acid structure.

- primary structure: linear sequence of the amino acids.

- secondary structure: when the polypeptide takes on a certain orientation in space. A coiling of the chain results in an α helix or a right-handed spiral.

- tertiary structure: final three dimensional shape. In muscles for example myosin molecules have a rod shape ending in globular heads.

- some proteins have only one polypeptide, while others have more than one polypeptide, each with his own primary, secondary, and tertiary structures. This arrangement gives a quaternary structure.

The final shape of a protein is very important to its function. Enzymes cannot function unless they have their usual shape. Under heat or extreme pH values, proteins undergo an irreversible change in shape, called denaturation. Some other points to our information: heating causes an egg white because albumin coagulates. Once a proteins loses its normal shape, it is no longer able to perform its usual function.

What are proteins? What is amino acid?

1.2.5   Nucleic Acids

Exist as two types:

- DNA, deoxyribonucleic acid (Fig. 1.13),

- RNA, ribonucleic acid.

Both these acids are polymers of nucleotides. Every nucleotide is a complex of three types of subunit molecules:

- phosphate (phosphoric acid)

- pentose sugar

- nitrogen-containing base.

The nucleotides in DNA contain the sugar deoxyribose, whereas, the nucleotides in RNA contain the sugar ribose. There are four different types of bases in DNA:

1. Adenine, A

2. Thymine, T

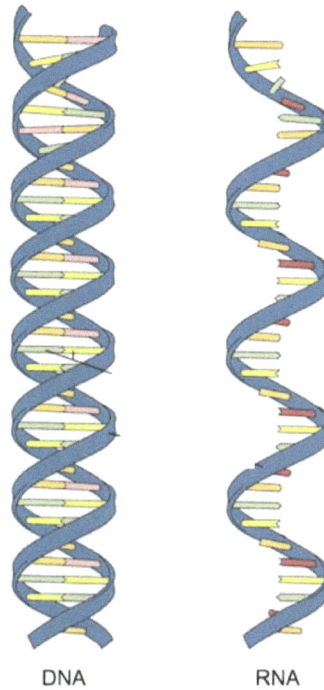

Figure 1.13: Comparison of DNA and RNA.

3. Guanine, G

4. Cytosine, C

In RNA, the base Uracil replaces the base thymine.

The nucleotides form a linear strand of molecules. The backbone is made of phosphate-sugar-phosphate while the bases project to one side of the backbone. The sequence of all the bases in human DNA, the human genome, assists in gene therapy, genetic counseling, and in the treatment of human illnesses.

DNA is double stranded. The two strands are twisted about each other in the form of a double helix. The two strands are held together by hydrogen bonds between the bases. When unwound, DNA resembles a stepladder.

- sides of the ladder: made up of phosphate and sugar molecules

- rungs of the ladder: complimentary paired bases T always with A and G always with C.

This complimentary base pairing allows DNA to replicate and ensure that the sequence of bases will remain the same. This sequence contains a code.

RNA is single stranded.

1.2.6 ATP

Adenosine is composed of adenine plus ribose. When Adenosine is modified by the addition of three phosphate groups gives ATP, the adenosine triphosphate. This is an energy carrier in cells which converts the energy of glucose is converted to that of ATP molecules. This ATP can be used to supply energy for chemical reactions in cells. By hydrolyzing the phosphate bond ATP is converted to ADP the adenosine diphosphate. This energy is used by the cell to synthesize macromolecules such as carbohydrates and proteins. In muscle cells, the energy is used for muscle contraction, while in nerve cells it is used for the conduction of nerve impulses.

Activities

1. Make a flow diagram showing how an ecosystem in a simple aquarium might work. Discuss the weakest parts in it!

2. Discuss the importance of external energy for the formation of life.

3. Discuss reasons why life on Earth is based on carbon and water. Could we expect that life elsewhere could have formed the same way? If yes give arguments for that.

4. What are the advantages of DNA based life?

5. Discuss the influence of *e.g.* UV radiation outbursts of the Sun on different mechanisms of cell division. Which kind of life will be affected stronger? Could that also be an advantage?

Send Orders of Reprints at bspsaif@emirates.net.ae

CHAPTER 2

Life on Earth

Abstract: This chapter reviews the origin and evolution of life on Earth. The collapse of an interstellar cloud into a protostar surrounded by a protoplanetary disk is reviewed. The early Earth was extreme hostile to life and first signs of life data from the time 3.8 billion years ago. Abiogenesis is discussed and it is highlighted that the process of evolution was not linear but it came about in several big steps. The idea of panspermia is reviewed. The famous Urey Miller experiment and other experiments are reviewed. Most probably life has started in deep undersea geysers called black smokers. During the last 500 million years, at least five times mass extinction has occurred. Possible astrophysical explanation for this extinction is given. The main message of this chapter is that we still do not completely understand abiogenesis though different possibilities are being discussed such as the synthesis of amino acids in Urey-Miller like environments, or near deep underwater geysers *etc.*

Keywords: Life; origin of life ; mass extinction; Earth: evolution of life; Earth: Origin; protoplanetary disk; early Earth; panspermia hypothesis; black smokers

2.1   Origin of Planet Earth

In order to understand origin of life on other planets or even satellites of planets, we must understand how life began on Earth. To our knowledge up to now, this is the only case where life has evolved definitely.

2.1.1   The Formation of the Stars

About 4.6 billion years ago a nebula consisting of gas and dust started to become gravitationally unstable. Gravity became stronger than the forces defined by the gas pressure and the nebula started to contract. This is also called Jeans instability, thereby a mass greater than the Jeans mass M_J becomes gravitationally unstable:

$$M_J = \left(\frac{\pi k_B T}{G \mu m_p}\right)^{3/2} \frac{1}{\rho_0^{1/2}} \tag{2.1}$$

Where, m_p is the proton mass, μ the mean molecular weight, k_B the Boltzmann constant, T the temperature of the gas. This equation can be derived from the dynamics of a gas which is given by the two equations:

$$\rho\left(\frac{\partial \mathbf{v}}{\partial t} + (\mathbf{v}.\nabla)\mathbf{v}\right) = -\nabla P - \rho \nabla \Phi \tag{2.2}$$

Figure 2.1: Artist's conception how stars from from an interstellar cloud.NASA

$$\frac{\partial \rho}{\partial t} + \mathbf{v}.(\nabla \rho) = -\rho \nabla.\mathbf{v} \tag{2.3}$$

the first equation gives the force balance, P is the gas pressure, Φ the gravitational potential. The second equation is the continuity equation, where $\mathbf{v}$ is the gas velocity field. The sound speed is given by

$$\nabla \mathbf{P} = c_s^2 \nabla \rho \tag{2.4}$$

From these equations the Jeans criterion can be derived by assuming that velocity and density can be written as two parts: (i) spatially uniform (subscript 0) and spatially varying (subscript 1). Then

$$\mathbf{v} = \mathbf{v_0} + \mathbf{v_1} \qquad \rho = \rho_0 + \rho_1$$
$$\frac{\partial \mathbf{v_0}}{\partial t} = \frac{\partial \rho_0}{\partial t} = 0$$

From this we obtain the linear equation in spatially varying quantities as

$$\frac{\partial \mathbf{v_1}}{\partial t} + \mathbf{v_0}.\nabla \mathbf{v_1} = -\nabla \Phi_1 - c_s^2 \nabla \left(\frac{\rho_1}{\rho_0} \right) \tag{2.5}$$

The Jeans instability determines when the star formation occurs in a molecular cloud. Under certain conditions, Jeans instability also gives rise to fragmentation. Typical values in interstellar clouds are:

Sound velocity: $c_s = 200\,\text{m/s}$
Density: $\rho_0 = 2 \times 10^{-20}\,\text{gcm}^{-3}$
Jeans wavelength: $\lambda = 1.7 \times 10^{18}\,\text{cm} = 0.56\,\text{pc}$.
Jeans mass: 27 $M_\odot$.

To form a Solar System, the initial molecular cloud containing about 30 solar masses must have fragmented into smaller clouds. Therefore, the sun was formed together with other stars, however we do not know where their location as they have dispersed into the galactic environment since their birth (see Fig. 2.1).

The collapse of the interstellar cloud occurred in different steps:

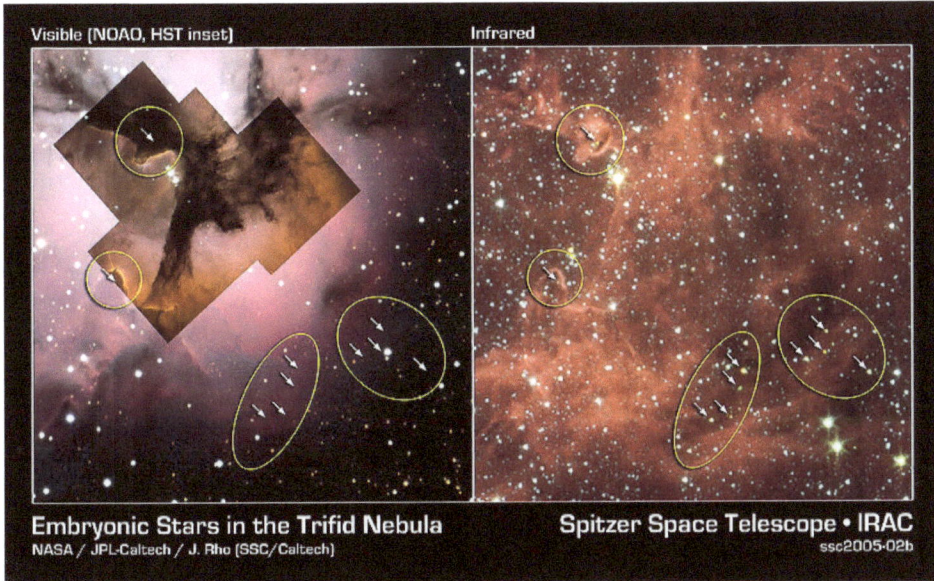

Visible [NOAO, HST inset]   Infrared

Embryonic Stars in the Trifid Nebula   **Spitzer Space Telescope • IRAC**
NASA / JPL-Caltech / J. Rho (SSC/Caltech)   ssc2005-02b

Figure 2.2: This image composite compares visible-light (taken with the Hubble Space Telescope) and infrared views from NASA's Spitzer Space Telescope of the glowing Trifid Nebula, a giant star-forming cloud of gas and dust located 5,400 light-years away in the constellation Sagittarius. Data of this same region from the Institute for Radioastronomy millimeter telescope in Spain revealed four dense knots, or cores, of dust (outlined by yellow circles), which are "incubators" for embryonic stars.

- The gas cloud became gravitationally unstable (Jeans criterion); at this stage it contained more than 30 solar masses.

- Trough fragmentation of the collapsing cloud during the adiabatic collapse, the so called Bok globules were formed.

- The collapse of the fragments continued, the density increased; gravitational energy got converted into heat, and the protostar started to glow in the infrared.

- T Tauri stars form amongst other stars. The mass of T Tauri stars is less than 2 solar masses and they emit jets of gas along their axis of rotation. Our Sun also was for several million years in such a T Tauri phase.

- A sun like star takes about 100 million years to form.

- A protoplanetary disk remained around the young star and small dust grains coalesced and formed larger planetesimals; this process lasted several 100 000 years.

In Fig.2.2 images from different telescopes are combined. The left image was taken with the Hubble[1] Space telescope (HST) which is a 2.4 m diameter telescope, placed in the Earth's orbit at a height of about 580 km. It was launched in 1990 with the NASA's space

---

[1]Named after E. Hubble, 1889-1953; discovered the expansion of the universe.

shuttle mission discovery. The Spitzer[2] Space telescope was launched on August 25, 2003. Because it mainly observes in the Infrared, the detectors have to be cooled down to -271°C. After five years, the cooling agent was consumed up and the telescope continued operating at -242°C. The main mirror has a diameter of 0.85 m and is made of Beryllium.

2.1.2   Protoplanetary Disks

Around a newly formed star a rotating protoplanetary disk can be observed. They can extend up to 1000 AU and the temperatures vary from 1000 K at the inner edge closest to the host star to several tens of Kelvin at the outer edge. The temperature can easily exceed 400 K inside 5 AU and 1,000 K inside 1 AU [33]. The heating of the disk is primarily caused by the viscous dissipation of turbulence in it and by the infall of the gas from the nebula. Such protoplanetary disks can be detected by spectroscopic or photometric observations. Spectroscopic signatures of dust particles can be detected in IR spectra of young stars. Matter in the protoplanetary disk falls into the central star, therefore the protoplanetary is an accretion disk. It cools and dust grains are being formed. In the outer regions of protoplanetary disks, at larger distances from the central star also ice grains are formed. These grains coalesce into km sized planetesimals. The formation of Moon to Mars-sized bodies may have taken only several 10^5 years.

The formation of the Earth lasted for about 500 million years and during this time, the formation of life was impossible because the surface of Earth was extremely hot, the atmosphere was hostile and the Earth was heavily bombarded by the residual planetesimals and objects in the solar cloud the most catastrophic event having been the formation of the Moon by an impact of a Mars-sized protoplanet on Earth.

The formation of the giant planets is less well understood.

Protoplanetary disks can be only observed around very young stars (an example is given in Fig. 2.3) so they must disappear after several tens of million years. The inner part of a disk is either accreted by the star of emitted in bipolar jets. The outer part can evaporate under the young star's strong UV radiation. Also the radiation pressure plays an important factor in the evaporation of dust particles.

The hypothesis of planetesimals is strongly supported by the observation of asteroids, comets and meteorites in the Solar System those being regarded as remnants.

2.1.3   Panspermia Hypothesis

After having discussed the formation of Earth we proceed discussing the formation of life on early Earth.

Did life really originate on Earth or was it transported from another object throughout space to earth? We know there exist organic compounds on meteorites and [43] show that organic compounds in a meteorite might also have formed in the early stages of the solar nebula.

The panspermia hypothesis suggests that life exists throughout the universe and it is transported and proliferated to all habitable objects by meteorites and comets. This requires that life survives the extreme conditions in space: (i) extreme cold (down to temperatures near absolute zero), (ii) hostile radiative environment in the vicinity of stars, (iii) extreme particle environment (cosmic radiation, high energetic particles), (iv) no nutrients over very long time spans. In the next chapter we will discuss extremophiles, some of which these harsh conditions for rather short time these harsh conditions. It is assumed that bacteria

---

[2]Named after L. Spitzer, 1914-1997; published important contributions to the physics of the interstellar medium.

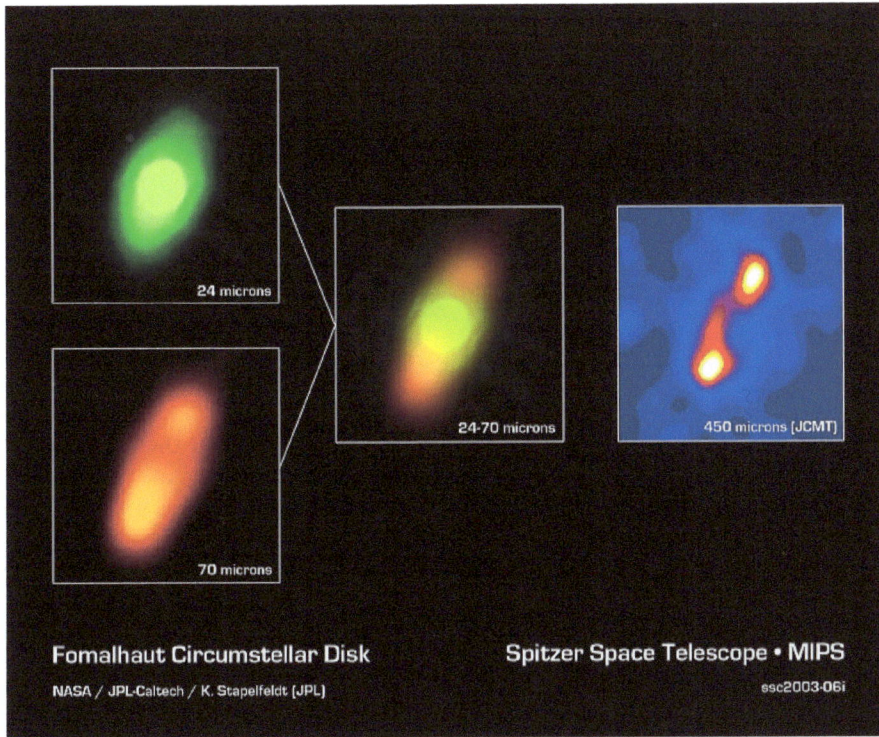

Figure 2.3: Fomalhaut and its protoplanetary disk. The 70-micron data (red) clearly shows an asymmetry in the dust distribution, with the southern lobe one-third brighter than the northern. Such an unbalanced structure could be produced by a collision between moderate-sized asteroids in the recent past (releasing a localized cloud of dust) or by the steering effects of ring particles by the gravitational influence of an unseen planet. At 24 microns (green), the Spitzer Telescope image shows that the center of the ring is not empty. [Note that an image of a reference star was subtracted from the Fomalhaut image to reveal the faint disc emission.] Instead, the 'doughnut hole' is filled with warmer dust that extends inward to within at least 10 astronomical units of the parent star. This warm inner disc of dust occupies the region that is most likely to be occupied by planets and may be analogous to our Solar System's 'zodiacal cloud' – but with considerably more dust. One possible explanation for this warmer dust is that comets are being nudged out of the circumstellar ring by the gravitational influence of massive planets. These comets then loop in toward the central star, releasing dust particles just as comets do in our own Solar System. Credit: NASA/JPL-Caltech/K. Stapelfeldt (JPL), James Clerk Maxwell Telescope.

might have travelled long distances in a dormant state and became active when exposed to habitable conditions on surfaces of planets. Then evolution started on these planets.

In the fifth century BC, the Greek philosopher Anaxagoras first mentioned this idea. It was later discussed by Berzelius, Kelvin (1871), Helmholtz (1879) and Svante Arrhenius (1903). Sir Fred Hoyle (1915-2001) and Chandra Wickramasinghe (born 1939) claimed that certain lifeforms may be responsible for epidemic outbreaks, new diseases when entering the Earth's atmosphere. In that context the term Macroevolution is used.

Panspermia occurs either interstellar (between stars) or interplanetary (between planets). Different transport mechanisms are suggested:

- Radiation pressure; this was first discussed by Arrhenius. Electromagnetic radiation emitted from stars exerts a pressure upon any surface exposed to it. Light consists of photons which are massless particles. Their energy E is related to their momentum p by the formula:

$$p = \frac{E}{c} \tag{2.6}$$

  Consider a beam of light that is absorbed on a surface. Then the momentum of the photons of that beam is transferred to the surface, thus transforming the momentum to it. Radiation pressure is about 10^{-5} Pa at Earth's distance from the Sun and decreases by the square of the distance from the Sun. The tiny particles in the dust tail of comets are influenced by the solar radiation pressure and point always away from the Sun.

- Lithopanspermia (microorganisms in rocks). This was suggested by Lord Kelvin in 1871.

- Spores that are trapped in plasmoid magnetic fields ejected from the magnetosphere.

- Directed pan spermia: a technologically advanced civilization sends spaceships with microorganisms throughout the galaxy.

A formula can be derived for the velocity of a grain particle of size r, density ρ and distance R_0 from a central object with luminosity L.

$$v = \sqrt{\frac{3L}{8\pi r \rho R_0}} \tag{2.7}$$

Let us consider a silicate grain in space at the distance of 1 AU[3]: $r \sim 10^{-5}$ cm, $\rho = 3 \, \text{gcm}^{-3}$, $R_0 = 1$ AU and $L = 4 \times 10^{33}$ erg/s [4]. Then this grain will obtain a velocity of 60 km/s away from the Sun this being larger than the gravitational escape velocity from the Solar System of an object near Earth's orbit (45 km/s). However, for larger particles the gravitational force will be more important than the radiative force. The radiative force at a distance R from a star with luminosity L is given by:

$$F_{\text{rad}} = \frac{L\sigma}{4\pi R^2 c} \tag{2.8}$$

let us consider a balance between the two forces, a particle with mass m is attracted by a star with mass M, then

---

[3] Astronomical Unit, mean Distance Earth-Sun
[4] Luminosity of the Sun

$$\frac{L\sigma}{4\pi R^2 c} = \frac{GMm}{R^2} \tag{2.9}$$

σ is the constant in Setfan Boltzmann's law. Therefore, we find that as soon as the mass of the grain exceeds a certain value, gravity will be stronger than the radiative force. The mass of a particle is $m = 4/3\pi r^3 \rho$, and:

$$r\rho = \frac{3L}{16\pi c GM} \sim 5.7 \times 10^{-5} \tag{2.10}$$

For a silicate grain $\rho = 3\,\mathrm{gcm}^{-3}$ this gives $r \sim 2 \times 10^{-5}\,\mathrm{cm}$. For larger grains, gravity is dominant. In this calculation two important effects have been neglected: (i) solar wind, (ii) cosmic rays. At Earth's distance the solar wind has a density of 1-10 particles per cubic centimetre (mainly protons). The speed is 400-600 km/s. The acceleration on a $10^{-5}\,\mathrm{cm}$ grain is 1000 times less than the acceleration due to radiation pressure. A similar result is obtained for the cosmic rays. Only in the case of early T-Tauri stars, the stellar wind could dominate the radiation pressure. There is one further mechanism. Dust grains at a size of $10^{-5}\,\mathrm{cm}$ might become electrically charged. Then they become coupled to the solar wind, which means they get velocities of up to 600 km/s. Proto-stars and protoplanetary disks have winds which had been more intense and had higher velocities than the solar wind, and the dust entrained in these winds has the same velocity as the winds. Therefore, radiation pressure only has minor contribution.

Interplanetary transport of material is well known. Meteorites of Martian origin (Fig. 2.4) or asteroid origin are found on Earth. In order to avoid pollution in our Solar System by manmade space probes, the space agencies implemented a sterilization procedure.

Plants, algae, fungi and some protozoans produce spores as part of the normal life cycle, which are highly resilient to UV and gamma radiation, desiccation, temperature, and starvation. They germinate when favorable conditions are restored. Bacterial spores have been found on Earth being 40 million years old. The DNA of some bacteria frozen in Antarctic glaciers has a half life of 1.1 million years. Given these facts, interplanetary panspermia could be possible. However, it is highly questionable whether the extreme long distance between stars that require a long exposure time would not mean a limit of interstellar panspermia. The distance between Sun-Earth corresponds to a light travel time of about 8 minutes, while the distance to Alpha Centauri is about 4.3 light years.

Compare the distance of the nearest star Alpha Centauri to the distance Sun-Earth! What half life for the DNA of simple living forms would be required for the interstellar transport between our Solar System and the Alpha Centauri system?

Finding any evidence of extraterrestrial life similar to ours would support the idea of exogenesis.

In the paper [110] a review on the panspermia hypothesis is given with additional literature.

2.1.4   Urey-Miller Experiment

Regardless whether life originated on Earth or was brought to Earth from some other place, it has to be explained how life could arise from inorganic matter through natural processes. Abiogenesis is the study of these complex reactions and processes.

Figure 2.4: Of over 53000 meteorites that have been found on Earth, 99 are martian. The Figure shows a meteorite from Mars found at Allan Hills of Antarctica (ALH 84001). Ejection from Mars seems to have taken place about 16 million years ago. Arrival on Earth was about 13 000 years ago. Cracks in the rock appear to have filled with carbonate materials.

Modern literature is full of articles on these topics, see *e.g.* [78]

We start with the Urey-Miller experiment which was conducted in 1952 and is still considered to be a classic experiment for the origin of life. It was published in 1953 by Stanley Miller (Fig. 2.5) and Harold Urey at the University of Chicago. The experiment used the following chemicals:

- water, H_2O

- methane, CH_4

- ammonia, NH_3

- hydrogen, H_2

These chemicals were sealed inside a sterile array of glass tubes, connected to a flask of liquid water and another flask containing electrodes. The water was heated and water vapor formed. Lightning was simulated by the electrodes. After some time the atmosphere was cooled again, and the water could condense. Within a day the mixture turned pink in color, at the end of one week, 10-15% of the carbon within the system was in the form of organic compounds and 2% had formed amino acids that make proteins in living cells, with the most abundant being glycine. Also sugars and lipids formed. In all experiments both left-handed and right-handed optical isomers were created in a racemic mixture. Note that in biological systems most of the components are non-racemic, or homochiral.

About four billion years ago, major volcanic eruptions occurred on Earth and they released CO_2, N_2, H_2S, SO_2 into the atmosphere. If these gases were included into the Urey-Miller experiment, more diverse molecules could appear. The early Earth atmosphere could have contained up to 40% of hydrogen. This would have been even more favorable for the formation of prebiotic organic molecules. Of course hydrogen escaped from the Earth's

Figure 2.5: Miller with part of the famous Urey-Miller experiment. NASA.

atmosphere because of its low molecular weight and relatively high atmospheric temperature. The velocity of atoms or molecules (weight μ) in an atmosphere is dependent on the temperature and molecular weight:

$$v = \sqrt{3kT/\mu} \qquad (2.11)$$

If the velocity is larger than the escape velocity $v > v_{\text{esc}}$

$$v_{\text{esc}} = \sqrt{2GM/R} \qquad (2.12)$$

where M is the mass of Earth and R its radius, the atom or molecule escapes.

There are several places in the Solar System and maybe in extrasolar planetary systems, where conditions similar to those of the Urey-Miller experiments are present however, the original Urey-Miller experiment is increasingly disfavored as an explanation of the origin of Earth life because the primodial atmosphere is probably not that reducing, but still mainly composed of CO_2, H_2O.

In many cases UV radiation may substitute the effect of lightning as energy source for the chemical reactions. One example is the Murchison meteorite which fell near Murchison, Australia in 1969. Being in space environment, this object was exposed to intense solar UV radiation. Over 90 different amino acids were found there. Similar processes as in the Urey-Miller experiments formed large amounts of complex organic compounds on the surfaces of comets and other objects. Since the early Earth was bombarded by comets, these complex compounds were brought to Earth.

2.1.5   Black Smokers

The Earth is geologically active. Tectonic plates are moving apart and on ocean basins, hydrothermal vents are found near fissures from which hot water escapes. On land we observe them as hot springs, fumaroles and geysers. Under the sea these hydrothermal vents may form the so called black smokers. Black smokers are similar to geysers on the surface but they are found on the floors of the oceans and they appear black because of the exhaustion of mineral rich warm or hot water into the cold ocean water minerals condensating out appearing as black.

Active hydrothermal vents are believed to exist on Jupiter's moon Europa and other satellites of Jupiter or even Saturn (Enceladus), and ancient hydrothermal vents have been speculated to exist on Mars.

On Earth, hydrothermal vents occur in the deep ocean along the Mid-ocean ridges: the East Pacific Rise and the Mid-Atlantic Ridge (Fig. 2.6). At these places, tectonic plates move away from each other and new crust is formed. At these depths the ambient water has a temperature of about 2^0C. Water that emerges from the vents has a temperature ranging from 60^0 to more than 400^0C. The pressure is about 218 atmospheres. Pure water can exist up to 375^0 in liquid form. At a depth of 3000 m the hydrostatic pressure is more than 300 atmospheres- salt water being denser than pure water.

The length of the world ridge-crest system is about 55 000 km. About 10 hydrothermally active sites have been visited and their fluids sampled by submersible.

Near Ascension Island on the Mid-Atlantic Ridge there exist three black smokers which are active since an earthquake occurred there in 2002. The measured temperatures of the water were up to 400^0C.

Minerals are deposited through these black smokers producing chimney like structures. Such a chimney could grow on the order of 30 cm per day. First the mineral anhydrite is deposited, followed by then sulfides of copper, iron and zinc.

Black smokers were first discovered in 1977 on the East Pacific Rise using a deep submergence vehicle called Alvin. White smokers are vents that emit lighter-hued minerals, such as those containing barium, calcium, and silicon. These vents also tend to have lower temperature plumes. An example of such a hydrothermal vent is given in Fig. 2.7.

Why are these hydrothermal vents important for the origin of life? Compared to the surrounding sea floor hydrothermal vent zones have a density of organisms 10,000 to 100,000 times greater. The organisms found there depend on chemosynthetic bacteria for food. Black smokers are the centre of entire ecosystems. There is no sunlight at these depths. Archaea and extremophiles (see next chapter) convert the heat, methane and sulfur compounds into energy by chemosynthesis. They are at the base of the food chain. Tube worms form an important part of the community around a hydrothermal vent. They have no mouth or digestive tract, and like parasitic worms, absorb nutrients produced by the bacteria in their tissues.

Long-term seafloor observatories will allow exploration of linkages between volcanism and this newly discovered biosphere in the vicinity of black smokers. Such approaches may provide essential new information about our own planet while providing critically needed insights into how we can explore other planets for life [48].

Wchtershuser ([108], [109]) proposed the Iron-Sulfur-World theory according to which life might have originated near hydrothermal vents.

More information about black smoker and chimney chemistry can be found in [106]. Growth of 'black smoker' bacteria at temperatures of at least 250^0C was investigated in [2].

2.2   Evolution of Life on Earth

2.2.1   Life: Autotrophic *vs.* Heterotrophic

An organism that produces complex organic compounds (such as carbohydrates, fats, and proteins) from simple inorganic molecules is called an autotroph. This requires energy which comes from light (photosynthesis) or chemical reactions (chemosynthesis). Autotrophs are the producers in the food chain, such as plants on land or algae in water. They do not use organic compounds as an energy source or a carbon source.

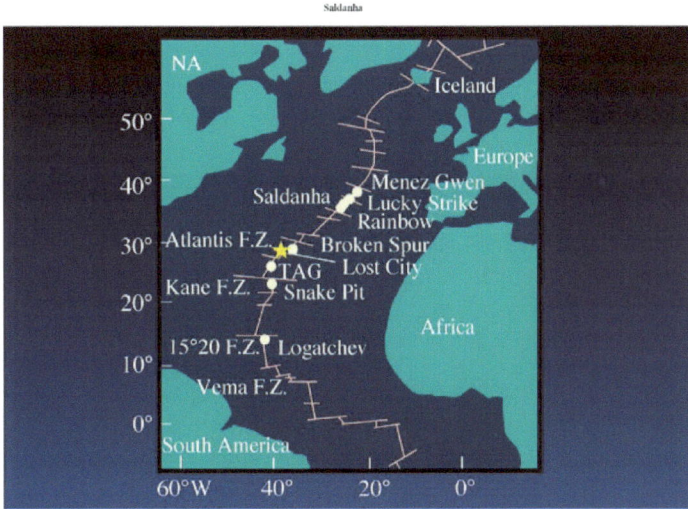

Figure 2.6: The Mid Atlantic Ridge. Credit: Univ.of Washington

Figure 2.7: A 5-foot-wide flange, or ledge, on the side of a chimney in the Lost City Field is topped with dendritic carbonate growths that form when mineral-rich vent fluids seep through the flange and come into contact with the cold seawater. Credit: National Science Foundation (University of Washington/Woods Hole Oceanographic Institution).

- phototrophs: use light as an energy source. The main reaction for photosynthesis is:

$$6CO_2 + 6H_2O \rightarrow C_6H_{12}O_6 + 6O_2 \tag{2.13}$$

$C_6H_12O_6$ is sugar. Light energy is gathered by chlorophylls. A part of it is stored in the form of ATP (adenosine triphosphate). The rest of the energy is used to remove electrons from a substance such as water. These electrons are then used in the reactions that turn carbon dioxide into organic compounds. The first photosynthetic organisms evolved about 3.5 billion years ago. They used other sources for the electrons than water, most probably hydrogen or hydrogen sulfide. The cyanobacteria appeared later as they started to enrich the Earth's atmosphere with oxygen. The efficiency of the photosynthesis is between 3% and 6%. By comparison, solar panels convert light into electric energy at an efficiency of approximately 6%-20%.

- lithotrophs or chemautotrophs: oxidize inorganic compounds such as hydrogen sulfide, sulfur, ammonium and ferrous iron. Most are bacteria or archaea that live in deep sea vents. There are several groups of chemoautorophs: methanogens, halophiles, sulfur oxidizers and reducers, nitrifiers, anamomox bacteria and others. In deep ocean iron oxidizing bacteria occur. They oxidze Fe(II) to Fe (III). There exist also manganese oxidizing bacteria.

They use ATP to reduce $NADP^+$ to NADPH to form organic compounds.

An organism that cannot fix carbon and uses organic carbon for growth is called heterotroph. They can be divided into

- photoheterotrophs: examples are purple bacteria, green bacteria. They produce ATP from light and use organic compounds to build structures.

- chemoheterotrophs: there exist two groups

  - chemoorganoheterotrophs: they use carbon compounds as energy sources, such as carbohydrates, fats and proteins from plants and animals.
  - chemolithoheterotrophs: use inorganic substances to produce ATP (*e.g.* sulfate reducing bacteria) and use organic compounds to build structure.

All animals and fungi are heterotrophic. Most protists and prokaryotes are also heterotrophic. Some form symbiotic relationships with autotrophs and obtain organic carbon from these symbiosis (*e.g.* corals). The biosphere can be divided into autotrophic and heterotrophic life.

The fundamental question is: was the first organism an autotroph (capable of synthesizing all its carbon compounds from CO_2) or was it a heterotroph? Wchtershuser ([109]) claimed that the source for a chemautotrophic origin of life was provided by the formation of pyrite in the following process

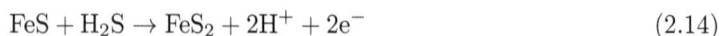

$$FeS + H_2S \rightarrow FeS_2 + 2H^+ + 2e^- \tag{2.14}$$

and the full reaction is given by

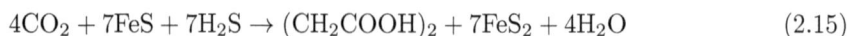

$$4CO_2 + 7FeS + 7H_2S \rightarrow (CH_2COOH)_2 + 7FeS_2 + 4H_2O \tag{2.15}$$

Let us now address the question of how life could have evolved from non living matter on Earth.

2.2.2   Abiogenesis

There are some basic facts about life on Earth:

- The first living organisms on Earth were single cell prokaryotes. Prokaryotes lack a cell nucleus. The oldest fossil microbe-like objects are dated to be 3.5 billion years old.

- Amino acids, often called the building blocks of life, can form via natural chemical reactions unrelated to life, as demonstrated in the Urey-Miller experiment and other similar experiments.

- Two important aspects of life have to be explained: replication and metabolism.

Aristotle believed in spontaneous generation of certain forms of life: mice come from dirty hay, crocodiles from rotting logs at the bottom of bodies of water *etc.* In 1665, Robert Hooke detected microorganisms, now called protozoa and bacteria. In 1668 Francesco Redi proved that no maggots appeared in meat when flies were prevented from laying eggs. The idea of biogenesis appeared, stating that every living thing came from a pre-existing thing (in Latin: omne vivum ex ovo). In 1871, Charles Darwin suggested that the original spark of life may have begun in a warm little pond. From that time on, the search for the origin of life moved to laboratory experiments. In 1924 Alexander Oparin discussed how "primeval soup" of organic molecules could be created in an oxygenless atmosphere through the action of sunlight. Later on other experiments such as the Urey-Miller experiment described earlier came up followed by many others.

There are two possible ways to put light onto the question how life originated:

- bottom-up-approach: try to synthesize a protocell using basic components.

- top-down-approach: manipulate prokaryotic cells with progressively fewer genes; where are the most minimal requirements for life?

The following stages have been proposed:

1. Origin of biological monomers; a monomer is an atom or small molecule that may bind chemically to other monomers to form a polymer (see Fig. 2.8). Amino acids are natural monomers that polymerize at ribosomes to form proteins. Nucleotides in the cell nucleus polymerize to form nucleic acids, DNA and RNA. Glucose monomers can polymerize to form starches, glycogen or cellulose.

2. Origin of biological polymers.

3. Evolution from molecules to cell.

All biologists now agree that bacterial cells cannot form from nonliving chemicals in one step. If life arises from nonliving chemicals, there must be intermediate forms, 'precellular life.' Some ideas here, for more details, consult the literature mentioned.

- DNA could succeed in starting life on its way. But even the shortest DNA strand needs proteins to help it replicate thus we arrive at a chicken-and-egg problem.

- "proteins first". Manfred Eigen said that because proteins form more easily they were first, but these first proteins must have been much shorter, otherwise it would have been very unlikely that they could have formed out of a soup of amino acids.

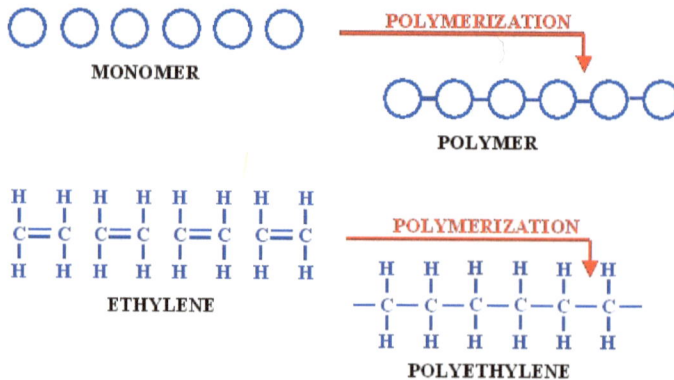

Figure 2.8: A simple example of a polymerization.

- Physicist Freeman Dyson[5] proposes to solve the chicken-and-egg problem with a double origin, one for metabolism (proteins) and one for replication (strands of nucleotides).

- A. G. Cairns-Smith[6] says that clay crystals could have served as the scaffolding upon which the first short DNA or RNA genome was constructed.

- Jeffrey L. Bada[7] assumes that the early Earth was frozen and believes precellular life started in "cold soup" under the ice.

- Claudia Huber and Gnter Wchtershuser say the soup where life originated was actually quite hot, probably near undersea volcanic vents, where iron and nickel sulfides might catalyze some of the necessary reactions.

- Thomas Gold[8] wonders if life might have originated in a hot environment even deeper, in Earth's crust.

- Stuart Kauffman[9] says,"...whenever a collection of molecules contains enough different kinds of molecules, a metabolism will crystallize from the brot" .

- Because starting the RNA world is so difficult, there probably needs to be a pre-RNA world. PNA, or peptide nucleic acid, might have some of the properties necessary to constitute that world. This would be pre-precellular life.

- In the 1970s Manfred Eigen[10] and Peter Schuster examined the transient stages between the molecular chaos and a self-replicating hypercycle in a prebiotic soup.

- The RNA world hypothesis describes an early Earth with self-replicating and catalytic RNA but no DNA or proteins.

---

[5] Freeman J. Dyson, Origins of Life, Cambridge University Press, 1985
[6] A. G. Cairns-Smith, Seven Clues to the Origin of Life, Cambridge University Press, 1985.
[7] Jeffrey L. Bada, "Cold Start," p 21-25, The Sciences May/June 1995
[8] Thomas Gold, "The Deep, Hot Biosphere", July 1992
[9] Stuart Kauffman, At Home in the Universe, Oxford University Press, 1995.
[10] Manfred Eigen, Steps Towards Life: A Perspective on Evolution (German edition, 1987), Oxford University Press, 1992.

From these statements one can see that it is still a hot debate how live really has originated. For our investigation about life in the universe it is interesting to see that there are quite different answers. Life could have originated in a relatively hot environment, or even cold environment.

In the paper of [57] the various prevailing hypotheses regarding origin of life-like abiogenesis, RNA world, iron-sulphur world and panspermia are examined, and the conclusion put forward is that delivery of life-bearing organic molecules by the comets in the early epoch of the earth alone possibly was not responsible for kick-starting the process of evolution of life on our planet.

2.2.3   RNA *versus* DNA World

Current life on Earth is being based on deoxyribonucleic acid (DNA), ribonucleic acid (RNA) and proteins. Both DNA and RNA can store genetic information, the double helix of the DNA gives it a greater stability. There exist several opinions as to whether RNA comprised the first self replicating system or was a derivative of an earlier system, like a pre-RNA[11].

The term RNA World was created by W. Gilbert in 1986. The hypothesis is strongly supported by the fact that most critical components of cells are composed mostly of RNA, so that in modern cells RNA appears as an evolutionary remnant of a no longer existing RNA world. Proteins as well as catalytic RNAs could function as catalysts in living matter.

Several properties of RNA could have been important for the origin of life:

- The ability to self-duplicate or duplicate other RNA molecules.

- The ability to catalyze simple chemical reactions; one example is a strand of RNA which creates more strands of RNA.

- The ability to catalyze the formation of peptide bonds resulting in short peptides or longer proteins.

The information storage in RNA is limited. Large RNA molecules can easily be broken down into their constituent nucleotides through hydrolysis. Therefore RNA based life becomes strongly prone to mutation.

The RNA world hypothesis starts with the existence of free-floating nulceotides in a primordial soup. The nucleotides formed bonds with one another. Being not stable these bonds often broke. However, certain sequences of base pairs stayed together, the chains grew faster than they were breaking down. These chains could be the first primitive forms of life. It was also speculated that viruses based on RNA were the last common ancestor of Bacteria, Archaea and Eukaryota. Some viruses evolved into DNA to protect their genes. Being common and abundant in the universe, the synthesis of RNA molecules could also be made by PAHs (polycyclic aromatic hydrocarbons). Also fullerenes could have played a role in this.

Being more stable than RNA, another assumption is PNA (peptide nucleic acid) could be a precursor of RNA.

The iron-sulfur theory proposes energy-producing cycles catalyzing the production of genes.

---

[11]See also *e.g.* the review The Roads to and from the world, Dworkin, J.P., Lazcano, A., Miller, S., Journal Theoret. Biology, 2003, 222,127-134

2.2.4   Prokaryotic, Eukaryotic Cells

There are two basic types of cells:

- Cells without nuclei and some other features: *prokaryotic* cells[12].

- *Eukaryotic* cells[13] possess a nucleus.

Most plant cells and animal cells are tiny and are invisible to the unaided eye. Prokaryotic and eukaryotic cells have in common the cell membranes, cytoplasm, genetic material, energy currency enzymes and coenzymes. All these are necessary to maintain a cell alive.

In biology, all organisms are classified into large categories called *domains*.

Most single-celled organisms are *prokaryotic bacteria*. There is another class that is called *archaebacteria*. Bacteria are a few micrometers long, have many different shapes such as spheres, rods or even spirals and are found on quite different habitats on Earth: in soil, acidic hot springs or even in radioactive waste and can survive outer space (that means extreme cold and vacuum). In a gram of soil, there are typically 40×10^6 bacteria and in a millilitre of water about 10^6. The ancestors of modern bacteria were single celled organisms that developed as first forms of life about 4×10^9 years ago. *Stromatolites* are fossils. They have been formed by sedimentation and cementation of microorganisms such as the cyanobacteria[14].

The *archaea* are more closely related to eukaryotes than the bacteria. Their genetic transcription and translation are similar to those of eukaryotes. Most bacteria and eukaryotes have membranes that are composed of glycerol-ester lipids. The membrane of the archaea is composed of glycerol-ether lipids. Archaea are found in extremely hot environments (hyperthermophiles) but they can be also found in all habitats.

2.2.5   Biological Classification

Classification is one of the fundamental aspects of science. The goals of biological classification are: assign names to organisms and group them according to their relatedness. This will provide us information on how these groups evolved. The study of evolutionary relationships is also called phylogenetics. If two organisms are related, this means that they share a common evolutionary history or phylogeny, both groups evolved from common ancestors.

A first naive classification was:

- plants (kingdom Plantae) and

- animals (kingdom Animalia).

Today, a system of five kingdoms is accepted which is shown in the Figure.

- Prokaryotes with the domain bacteria, archaea

- Eukaryotes with the domain eubacteria, kingdoms protista, fungi, plantae, animalia.

As it is shown in Fig.2.9, this represents also the evolution of the organisms. The animalia evolved last.

The Kingdom monera includes the prokaryotes and is divided into Eubacteria (true bacteria) and the Archaea. Just for illustration let us give the biological classification of the Homo sapiens: Domain: Eukaryota, Kingdom Animalia, Phy- lum Chordata, Class Mammalia, Order: Primates, Family: Hominidae, Genus: Homo, Species: sapiens.

---

[12]pro means before, karyon means nucleus
[13]eu means well or good.
[14]An old term no longer used now is blue green algae.

EUKARYOTES

PROKARYOTES

Kingdom
Animalia

Kingdom
Plantae

Kingdom
Fungi

Kingdom
Protista

Domain:
Bacteria

Domain
Archaea

Domain
Eukarya

Common ancestor

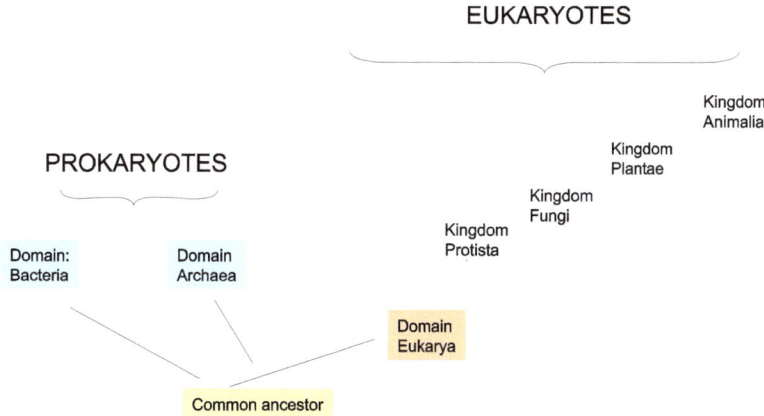

Figure 2.9: The major groups of living organisms on Earth.

2.2.6   Life and the Environment

Life on Earth first appeared about 3.5 Billion years ago. Since then, life considerably changed the environment on our planet. The cyanobacteria are also known as blue-green bacteria, blue-green algae. They can be found in oceans, fresh water or even in moistened rocks in deserts. Today, they account for 20-30% of Earth's photosynthetic productivity. Chloroplasts found in eukaryotes (algae and plants) likely evolved from an endosymbiotic relation with cyanobacteria. Stromatolites of fossilized oxygen-producing cyanobacteria have been found from 2.8 billion years ago.

Today, the Earth's atmosphere consists about 20% of oxygen. This oxygen was mainly produced by cyanobacteria. Oxygen transformed Earth's atmosphere to one suitable for the evolution of aerobic metabolism and complex life. The release of oxygen into the early Earth atmosphere is still not fully clear.

Organic matter becomes buried in seafloor sediments that later hardened into rocks. Cyanobacteria performed oxygenic photosynthesis and this latter process produced organic carbon and oxygen. When these organisms die, their remains become buried in seafloor sediment. This decomposition removes oxygen from seawater and also from the atmosphere. The carbo-burial rate remained constant however and cannot explain the buildup of oxygen in the atmosphere.

As the carbon-burial theory goes, when organic material is buried, oxygen becomes available to build up in the atmosphere. So perhaps there was a sudden increase 2.3 billion years ago in the amount of organic carbon that was buried, leaving more free oxygen. Tectonic and volcanic activity released gases. Gases like Carbon monoxide got reduced, so in the released gases there was more oxygen present than in the atmosphere. Thus a combination between photosynthetic production of oxygen together with the release of oxygen through volcanic activity could explain our oxygen rich atmosphere.

The early Earth's atmosphere was probably rich of methane. Methane being a strong greenhouse gas could have been responsible for a warming of the early Earth keeping it from a frozen state which would otherwise be the consequence of a less luminous early Sun[15]. Later, methane became photolyzed by the solar UV radiation.

---

[15] According to stellar evolution, young stars are more active but less luminous than older ones.

2.2.7   Mass Extinction

The evolution of life was not linear, several mass extinctions occurred. Complex organisms evolved from simple unicellular organisms, with some cells being specialized to specific tasks. Complex organisms need more energy to maintain their function, therefore a transition from photosynthesis to aerobic respiration was necessary since respiration is more efficient from the energetic point of view. However, such a transition became only possible as the Earth's atmosphere became enriched with oxygen. An atmosphere enriched with oxygen provides two functions:

- enables respiration to complex organisms The respiration process is mainly:

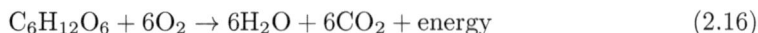

$$C_6H_{12}O_6 + 6O_2 \rightarrow 6H_2O + 6CO_2 + \text{energy} \qquad (2.16)$$

- provides an ozone layer that gives a protection against harmful UV radiation, therefore life on land could develop.

The quasi outburst of life on land occurred about 500 million years ago.

The prize for these evolutionary quantum leaps was a greater vulnerability to changing environmental conditions that result from cosmic catastrophes. We can state that:

- primitive life (bacteria, archaea): less vulnerable to cosmic catastrophes; once having formed on a planet, it is highly probable to survive a cosmic catastrophe.

- complex life: multicellular; more vulnerable to catastrophes; chances to sur- vive cosmic catastrophes are smaller. Needs a stable continuous habitable zone for a long time; cosmic catastrophes led to mass extinction and provided the chance to mutations that evolved to more complexity.

What are possible cosmic catastrophes?

- impacts: as it was already mentioned in the early phase of planet formation, in our Solar System impacts occurred frequently, the period being called cosmic bombardment. Two positive effects of such a phase of heavy collisions within a relatively short time could have been (i) impacts of comets delivered water to the planets, (ii) impacts delivered organic compounds such as amino acids. Today, the probability of a global catastrophe by an impact is low, maybe once every several 10^7 years. Mass extinctions during which a large fraction of life became extinct, occurred several times over the last 500 million years and at least some of them can be explained by an impact of an asteroid or comet. To cause a global catastrophe the impacting object must be larger than 1.5 km. The most famous impact, also called KT event, occurred 65 million years ago, and caused the extinction of the dinosaurs.

- solar outbursts: these will be discussed in more detail in the chapter about the Sun. Though our modern civilization is vulnerable to power failures, communication breakdowns, satellite damages *etc* which are all caused by large solar outbursts (flares, CMEs), a global catastrophe can be excluded.

- nearby supernova explosion: supernovae occur when massive stars are transformed to their final stages (neutron stars or even black holes) at the end of their evolution. During a supernova outburst large quantities of UV and gamma radiation are emitted which can destroy the ozone layer and change the atmosphere of planets. Since the

Solar System moves about the center of our Galaxy, the cosmic neighborhood changes and it seems probable that a nearby supernova explosion can occur once every one billion years.

From this list we conclude that the most probable cosmic catastrophes may come from impacts. Cosmic catastrophes were also positive for the evolution of life. Without the KT event, dinosaurs would still be the dominant lifeforms on Earth.

2.2.8   The Main Stages in the Evolution of Life on Earth

In this chapter we summarize the main stages which seem to be crucial for the evolution of life on Earth. If life appears elsewhere in a similar form then these stages might be also similar on such planets.

1. About 4.5 billion years ago: Formation of the Moon. As it will be discussed in a later chapter the Moon was formed by a collision of the Early Earth with a protoplanet. The Moon has a stabilizing effect on the Earth's rotation axis which stabilizes the climate on Earth and it is also assumed that tides played an important role for the transition of life from sea to land.

2. About 3.5 billion years ago: oldest microbial life fossils.

3. up to about 2 billion years ago: no free oxygen in Earth atmosphere.

4. About 2.7 billion years ago: Sterols = Eukaryotes[16]?

5. About 1.5 billion years ago: Eukaryotes appear.

6. 750 million years ago: multicellular eukaryotes.

7. 600 million years ago: Cambrian explosion.

The cause for the Cambrian explosion of life on Earth is still under debate. By the start of the Cambrian, the large supercontinent Gondwana, comprising all land on Earth, was breaking up into smaller land masses. This increased the area of continental shelf, produced shallow seas, thereby also expanding the diversity of environmental niches in which animals could specialize and speciate. The geology and habitability of terrestrial planets are reviewed in [98].

Activities

1. In physics, there exists the concept of Entropy. If you are not familiar with that please consult textbooks. Discuss the concept of entropy in the context of life. Could life be described as localized entropy decrease and if yes how can this be reconciled with the entropy increase law in thermodynamics?

2. In this chapter we discuss life as it appears on Earth. Give arguments that how life elsewhere could be similar to that on Earth. What are the minimum similarities to be expected?

3. Consider the entire Earth as an Ecosystem. What are the external factors on this system?

4. Discuss how complex life could have evolved from single celled organisms. Was this a logical process or unexpected?

---

[16]Sterols and related compounds play essential roles in the physiology of eukaryotic organisms.

Send Orders of Reprints at bspsaif@emirates.net.ae

CHAPTER 3

How Earth Protects Life

Abstract: After discussing the basic properties of life and origin on Earth (which is still debatable) we highlight the role played by earth in protecting life, especially complex life. The atmosphere protects life on the Earth's surface from short wavelength radiation, the Earth's magnetosphere provides a shielding against energetic charged particles. Life even exists under extreme conditions on Earth. We have already discussed black smokers in the context of origin of life and we will give further examples of life existing in at first sight hostile conditions. The study of extremophiles has enhanced the chance to find life elsewhere under extreme environments such as on the surface of Mars. The main message of this chapter is that life is being found to exist in environments that have been thought as hostile several decades ago which gives us hope that the extension of a habitable zone might be broader.

Keywords: Extremophiles; radiation dose; radiation limit; radiation shielding; Earth: atmosphere; Earth: magnetosphere; magnetic pole reversals;

3.1   Radiation and Energetic Particles Shielding

3.1.1   Energetic Radiation

The electromagnetic Spectrum

The electromagnetic spectrum covering all wavelengths, from short to long wavelengths, are mainly named as

- X-rays: further divided into hard and soft X-rays.

- UV radiation: further divided into far UV (FUV), and normal UV.

- Visible light, wavelength from 400 - 700 nm: our eyes are adapted to that wavelength range because the Earth's atmosphere is transparent to it and more than 90% of sunlight is emitted in that range.

- Infrared, IR.

- Radio waves.

In Table 3.1 the wavelengths are given for the different ranges.

From physics we know that the energy of light is directly proportional to its frequency ν or inversely proportional to its wavelength λ:

Table 3.1: Spectral ranges and their wavelengths.

Range	Wavelength	Daily use
Gamma Rays	around 1 pm	Medical
X Rays	around 1 nm	Medical
UV	0.001-0.4μm	Sun tan
Visible	0.4-0.7 μm	Seeing
Inrared, IR	0.7-10 μm	Physiotherapy
Microwaves	1 mm - 50 cm	Ovens
Television	around 50 cm	Television broadcats
Radio	1m - 1500 m	Radio broadcasting

$$E = h\nu = hc/\lambda \tag{3.1}$$

The shorter the wavelength will be the more energetic will be the radiation and the more harmful to complex molecules present in living cells.

How to measure radiation

The *rad* represents a certain dose of energy absorbed by 1 gram of tissue. The number of rads received would be the same, be it of a single cell, an organ (*e.g.*, an ovary) or the entire body[1].

The unit *rem* (Rntgen Equivalent Man) is defined by

$$rem = rad \times Q \tag{3.2}$$

Q is a quality factor depending on the type of radiation. For gamma rays and X rays $Q = 1$, neutrons have Q=5 and alpha particles Q=20. 100 rem = 1 Sv (Sieverts). Now the rad is often replaced now by the Gray, and 1 Gray = 100 rad.

For comparison: Dose from a typical set of full-body computed tomography (CT) scans is 45 mSv, while a typical chest X ray is 0.1 mSv. Approximate lethal dose ("LD50") if no treatment had been given to the entire body in a short period is 4,500 mSv. An acute radiation dose is defined as a large dose (10 rad or greater, to the whole body) delivered during a short period of time (most probably on the order of a few days). If large enough, it may result in effects which are observable within a period of hours to weeks. Radiation sickness is observed for acute dose $\sim$ 100 rad, about 50% of a population may die within several weeks for an acute dose > 450 rad.

A chronic dose is a relatively small amount of radiation received over a long period of time. The body is better equipped to cope with a chronic dose than an acute dose.

The average person in the United States receives about 360 mrem annually as whole body equivalent dose from background sources. A person in France receives about 2.4 mSv per year and a person in Kerala about 1 Sv per year.

The recommended dose limits for worker set by the US Nuclear Regulation Commission are listed in Table 3.2.

---

[1]The concentration of salt in sea water is the same considering one cm^3 or an entire ocean.

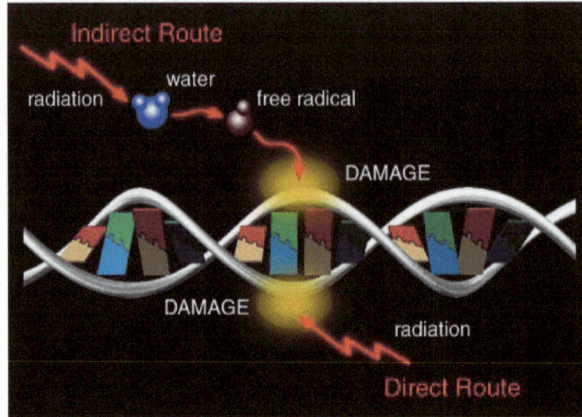

Figure 3.1: Direct and indirect routes of radiation damage on a DNA.

Table 3.2: Radiation dose limits given by NRC, the United States Nuclear Regulation Commission. TODE stands for Total Organ Dose Equivalent, TEDE for Total Effective Dose Equivalent, and LDE for eye dose equivalent (lens of an eye at a tissue depth of 0.3 cm). Sometimes, instead of rem the unit Sieverts, Sv is also used, *e.g.* 50 rem = 0.5 Sv.

Organ	Limit	Remarks
Whole Body	5,000 mrem/yr	TEDE
Any Organ	50,000 mrem/yr	TODE
Skin	50,000 mrem/yr	SDE
Extremity	50,000mrem/yr	SDE
Lens of Eye	15,000 mrem/yr	LDE
Embryo/Fetus	500 mrem/yr	
Member of the Public	100 mrem/yr	

Biological Effects

Biological effects of radiation on cells begin with the ionization of atoms. When the electron shared by the two atoms to form a molecular bond is dislodged by ionizing radiation, the bond is broken and thus, the molecule falls apart affecting different parts of a cell. The most critical parts are considered to be the chromosomes since they contain genetic information for a cell to function to reproduce. There are three possible steps:

- Cells are damaged but are able to repair themselves[2].

- When cells are damaged, repair mechanisms start to work but the cell functions abnormally. This malfunction of cells leads to cancer.

- Cells die.

Cells which divide rapidly and being not highly specialized are normally more prone to be affected by radiation. An example is blood forming cells.

The structure of the DNA can be destroyed. If a gamma ray passes through a cell, the water molecules near the DNA might be ionized and the ions might react with the DNA causing it to break. The strands can be broken by a direct route, and indirectly when a free radical acts. Single strand breaks can be repaired in most cases by the organism however double strand breaks suffer an ultimate damage since they can not be repaired. The direct and indirect routes of radiations causing damage to DNA are shown in Fig. 3.1

How can cells repair after radiation damage? Discuss the consequence of cell damage.

3.1.2 Energetic Particles

Energetic particles affecting the Earth have different origins. They may be part of the solar wind, a continuous flow of charged particles or of solar energetic events (flares and CMEs, will be discussed later) or they may be of cosmic origin arising from explosive events such as supernova explosions. The percentage of the contribution of different subatomic particles in cosmic rays is shown in Fig. 3.2.

About 10-15% of the natural background radiation energy of the Earth is contributed by cosmic rays. The other percentages come from natural radioactivity. As for radiation, the intensity of cosmic rays is much higher outside the Earth atmosphere and magnetosphere, thus providing a serious danger to manned space missions. The flux of incoming cosmic rays is dependent on

- Magnetic field of Earth,

- Solar wind,

- Energy of the particles.

When colliding with particles of the upper Earth atmosphere, they produce showers of secondary particles like pions, and kaons that decay into muons. By means of gamma ray spectroscopy, cosmic ray particles hitting the surfaces of planets can be detected, because the resulting gamma ray emissions are characterized by energies above 10 MeV being considerably higher than expected from natural radioactivity.

---

[2]Roughly, between 10 and 100 mSv DNA self repair mechanisms work.

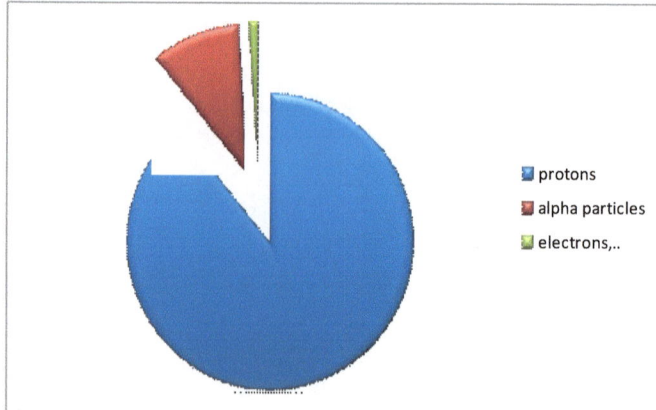

Figure 3.2: Percentage of different cosmic ray particles.

What are the sources of energetic particles arriving at the Earth?

3.1.3 The Earth Atmosphere

The Earth atmosphere provides a natural protection against high energetic radiation. The most famous protective zone is the Ozone Layer. This layer was detected in 1913 mainly concentrated in the lower stratosphere between 20 and 30 km above the surface of Earth. Ozone concentrations are greatest between about 20 and 40 km, where they range from about 2 to 8 parts per million. If all of the ozone was compressed to the pressure of the air at sea level, it would be only 3 millimeters thick. The formation of ozone was explained by Chapman in 1930. The simplified reaction is:

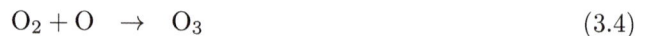

$$O_2 + h\nu \quad \rightarrow \quad O + O \tag{3.3}$$
$$O_2 + O \quad \rightarrow \quad O_3 \tag{3.4}$$

Here, $h\nu$ denotes an UV radiation quantum. UV radiation that is relevant for the formation of the ozone layer can be divided into three different wavelength ranges:

- UV-a: 400-315 nm

- UV-b: 315-280 nm

- UV-c: 280-100 nm

Being predominantly absorbed above 35 km (for wavelength $< 200\,\text{nm}$) in the ozone layer, UV-c would be extremely harmful to organisms. UV-b causes severe sunburns, and its excessive exposure may lead to skin cancer. UV-a reaches the surface and causes no damage in relatively short exposures. Stratospheric ozone is produced by solar UV radiation, its distribution over the globe being quite complicated.

The atmosphere absorbs also far UV and X rays by dissociation of various molecules like O_2. In Fig. 3.3 it is shown that how deep different kinds of radiation along with the particles of different energies can penetrate. Besides this absorbing feature, the dense Earth atmosphere also compensates for the day/night differences in temperature.

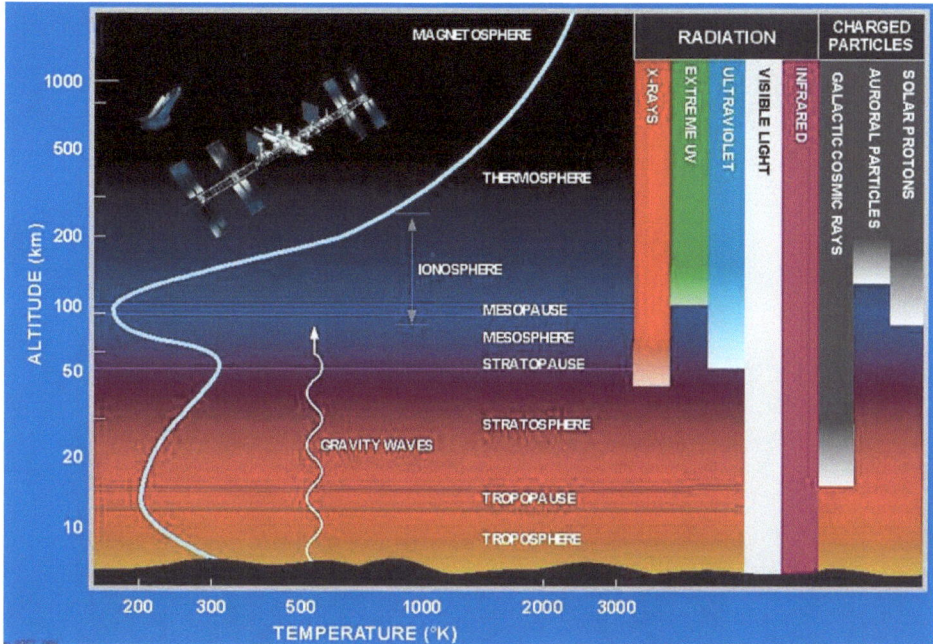

Figure 3.3: The Earth atmosphere. It has been shown how different parts of electromagnetic radiation are absorbed at different height.

3.1.4 The Earth Magnetosphere

The Earth having a magnetic field that can be approximated near the surface of the Earth by a dipole field (see Fig. 3.4). The dipole is located at the center of the Earth but tilted by 10^0 with respect to the rotational axis. The magnetic South pole is near the geographic North pole. The strength varies strongly as intensity is stronger near the poles and weaker near the equator ranging from 25000 to 60000 nT (0.25 to 0.60 Gauss). A strong refrigerator magnet has a field of about 100 Gauss. The minimum magnetic filed has been recorded to occur over South America while there are maxima over northern Canada, Siberia, and the coast of Antarctica south of Australia. The solar wind is responsible for the overall shape of Earth's magnetosphere at larger distance from the surface of Earth, and fluctuations in its speed, density, direction, and entrained magnetic field strongly affect Earth's local space environment. For example, the levels of ionizing radiation and radio interference can vary by factors of hundreds to thousands. Charged particles from the sun are trapped in the radiation belts and some manage to penetrate the magnetic shield and cause aurorae.

How can we explain the existence of the Earth magnetic field? The liquid outer core of the Earth consists mainly of molten iron being highly conductive. Loop currents cause a magnetic field, a changing magnetic field causes an electric field and both these electric and magnetic fields act on flowing currents. These are the fundamentals of dynamo theory to explain the existence of a magnetic field on a planet. Since the magnetic field is not constant, field reversals have occurred. Reversals occur at apparently random intervals ranging from less than 0.1 million years to as much as 50 million years. The most recent such event, named as Brunhes-Matuyama reversal, occurred about 780,000 years ago. There has not been any correlation found between mass extinctions and magnetic field reversals. During such reversals there remains still a residual field so the Earth is always protected, but the intensity of protection varies.

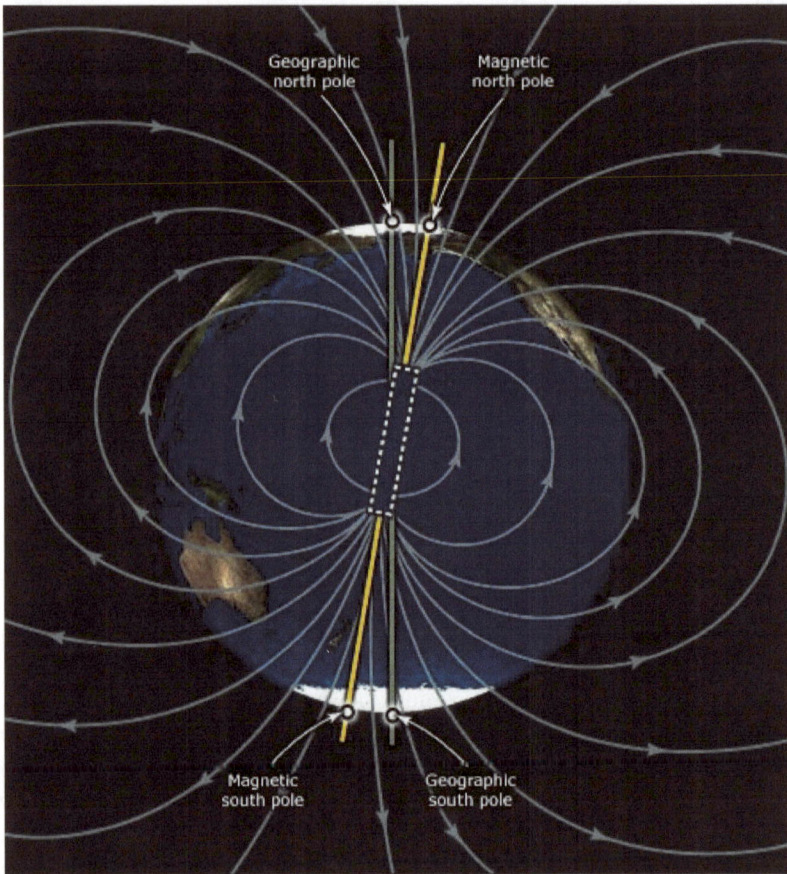

Figure 3.4: The magnetic field near Earth can be represented by a dipole field.

It has been claimed quite often, that during periods of magnetic pole reversals, mass extinction occurred on Earth. What would be your answer? Imagine a planet without a magnetic field. How it protected against cosmic ray particles and the particles emitted by its host star?

Summarizing we state that the magnetic field provides a shielding to the atmosphere against charged particles. Without having a magnetosphere, the surface of a planet would be bombarded by energetic charged particle coming mainly from its host star destroying cells of possible lifeforms as it was discussed above. Magnetospheres can be explained by a liquid core. If a planet is too small, it cooled and no liquid core remains. This happened to the Moon as well as in the case of Mars.

The atmosphere itself provides a shielding mainly against radiation.

3.2   Life Under Extreme Conditions

We have seen that the Earth provides two major protections against radiation and energetic charged particles: the atmosphere and the magnetosphere. However, both magnetosphere as well as atmosphere are not constant. There are reversals of the magnetic poles and the composition of the atmosphere changes considerably during the evolution of the Earth. Could life on Earth be resistant to such changes? How could life survive cosmic catastrophes, those discussed already?

An organism living in physically or geochemically extreme conditions being detrimental to most life on Earth is called an extremophile. The classification and description of these mainly microbic organisms are extremely difficult thus restricting us to give a few examples here.

Life under extreme environments, what it means to be an extremophile, and, the implications of this for evolution, biotechnology and especially the search for life in the Universe has been discussed by [89] and by [19].

In Table 3.3 a summary of extremophiles is listed.

3.2.1   Extreme Temperatures

Thermophiles are found in various geothermally heated regions of the Earth, such as hot springs like those in Yellowstone National Park (see image) and deep sea hydrothermal vents, as well as in decaying plant matter, such as peat bogs and compost. They can exist at temperatures between 45 and 122^0 C. Many of the hyperthermophiles require sulfur for growth. Some cause bright colors in hot water lakes (Fig. 3.5). Hyperthermophiles like temperatures above 80^0C. The most heat resistant of these microbes is Pyrolobus fumarii growing in the walls of smokers. It reproduces best at a temperature of about 105^0 C, and can multiply in temperatures up to 113^0 C. At temperatures below 90^0 C the hyperthermophile stops growing.

Is there an upper temperature limit for life? Before the discovery of thermophilic microbial life above 100^0 C, it was believed that 80^0 C was the upper limit for life. This idea was based upon the fact that at sea level, DNA unravels since being unable to maintain the double-helix structure at such temperatures. Today, scientists believe that the upper limit for life comes around 160 C. This is because at temperatures above 160 ^{0}C, ATP, being used by all living organisms for energy, begins to deteriorate rapidly.

The fact that there do exist such heat resistant organisms capable of withstanding temperatures of 160^0 C gives rise to the possibility for life in similar conditions elsewhere in the

Table 3.3: Extremophiles.

Name	Environment
Acidophiles	Live in pH $\leq$ 3
Alkaliphiles	Live in pH $\geq$ 9
Barophiles	Live under high pressure
Piezophiles	-"-
Endoliths	Live in microscopic spaces within rocks
Cryptoendoliths	Live in fissures, faults in deep subsurface
Halophiles	Need salt to grow
Hyperthermophiles	Live at T $= 80 - 120^0$C , hydrothermal vents
Hypoliths	Live inside rocks in cold deserts
Litoautotrophs	Source of carbon is CO_2
Metatolerants	tolerate high levels of metals like Cu, Cd, Zn.
Oligotrophs	Grow in nutritionally limited environment
Osmophiles	Grow in sugar concentration
Psychrophiles, cryophiles	Grow at T $< -15^0$C; permafrost, polar ice..
Radioresistants	Tolerate high levels of ionizing radiation
Thermophiles	Live at temperature up to 120^0C
Xerophiles	Grow in extremely dry conditions

Figure 3.5: Aerial view of Grand Prismatic Spring; Hot Springs, Midway & Lower Geyser Basin, Yellowstone National Park. The spring is approximately 250 by 300 feet (75 by 91 m) in size. This photo shows steam rising from hot and sterile deep azure blue water in the center surrounded by huge mats of brilliant orange algae and bacteria. The color of which is due to the ratio of chlorophyll to carotenoid molecules produced by the organisms. Courtesy: Jim Peaco, National Park Service.

universe. Such locations could include hydrothermal vents in the deep seas of Europa one of the satellites of Jupiter, or inside the volcanoes on Titan, Saturn's biggest moon.

On Earth, ideal places to study thermophiles are the Yellowstone National Park, Iceland and Kamchatka.

3.2.2   Extreme Chemical Environments

Acidophilic organisms live under extremely acidic conditions (pH< 2.0). In chemistry, pH is a measure of the acidity or basicity of an aqueous solution. Pure water is said to be neutral, with a pH close to 7.0. Solutions with a pH less than 7 are considered to be acidic and solutions with a pH greater than 7 as basic or alkaline. In a solution pH approximates but is not equal to p[H], the negative logarithm (base 10) of the molar concentration of dissolved hydronium ions, H_3O^+; a low pH indicates a high concentration of hydronium ions. To compensate such a high concentration of hydronium ions, most acidophile organisms have been evolved under extremely efficient mechanisms to pump protons out of the intracellular space in order to keep the cytoplasm at or near neutral pH.

Acidophiles

Acidophiles can be classified as Bacteria, Archaea and Eukaryotes. Some examples:

- Archaea: Sulfolobales, Thermoplasmatales, ARMAN, Acidianus brierleyi, A. infernus, facultatively anaerobic thermoacidophilic archaebacteria, Metallosphaera sedula, thermoacidophilic.

- Bacteria: Acidobacterium, Acidithiobacillales, an order of Proteobacteria *e.g.* A.ferrooxidans, Thiobacillus prosperus, T. acidophilus, T. organovorus, T. cuprinus, Acetobacter aceti, a bacterium that produces acetic acid (vinegar) from the oxidation of ethanol, Alicy-clobacillus, a genus of bacteria that can contaminate fruit juices.

3.2.3   Extreme Physical Environment

Bacteria including Escherichia coli and Paracoccus denitrificans were tested under conditions of extreme gravity. The bacteria were cultivated while being rotated in an ultracentrifuge at high speeds corresponding to 403,627 times "g" (the normal acceleration due to gravity). Paracoccus denitrificans was one of the bacteria which displayed not only survival but also robust cellular growth under these conditions of hyperacceleration usually found only in cosmic environments, such as on very massive stars or in the shock waves of supernovas.

Living cells are strongly influenced by pressure, changing their physiology and biochemistry. Barotolerant cultures tolerate high pressure, barophilic cultures have become dependent on high pressure. Deep sea bacteria show both types. Barotolerant bacteria under high hydrostatic pressure regulate the fluidity of membrane phospholipids to compensate for pressure gradients between the inside of the cell and the environment. Extreme barophiles grow at pressure larger than 700 atm and do grow at low pressures.

In the literature, many research articles are available about extremophiles. Damage avoidance and DNA repair mechanisms of extremophiles, particularly radiation resistance of the halophilic archaeon Halobacterium salinarum to ionizing radiation have been discussed in [86]. Extremophiles and chemotrophs as contributors to astrobiological signatures on Jupiter's moon Europa are reviewed by [96]. The role of extremophiles for search of life in the universe was stressed by [88], [67] and [19].

Activities

In the first three chapters of the book we have discussed the origin of life, its evolution and work. In the next chapters we will first describe the other bodies in the Solar System and then try to apply these concepts.

- What are the special mechanisms extremophiles have developed to survive their environment?

- Discuss how extremophiles could have evolved by selection process. Which life came into being first, the extremophiles or the life under normal conditions as we know it.

- What should be the upper and lower temperature limits for life?

- Is there any chance that life can proliferate under even more extreme temperatures than the temperature range defined by the extremophiles?

Send Orders of Reprints at bspsaif@emirates.net.ae

CHAPTER 4

The Solar System

Abstract: The Solar System can be regarded as a prototype for a planetary system. Many planetary systems have been detected over the last two decades, however, we will not be able in the near future to explore these exoplanets with such precision as the planets in our Solar System. From the details of our Solar System we can extrapolate to the exoplanets and the study of these systems helps us to better understand the formation of our own system. Parts of this chapter are from the book of the author [42]. We shall discuss the different classes of objects in the solar system such as the big planets, dwarf planets, asteroids and comets. First it will be outlined how we obtained the important physical parameters of the planets such as temperature, composition, mass, radius *etc.* from observations form the Earth. We will also discuss why Venus is such a dry planet and whether there is water on Mars. Then the objects will be discussed in more detail. Finally some interesting satellites and the rings of planets will be discussed. The main message of this chapter is that in the Solar System life could exist only on two categories of objects namely planets and some satellite of planets.

Keywords: Planets; Solar System; terrestrial planets; giant planets; Galilean Satellites; Titan; Europa; Ganymede; Io; Callisto; Mercury; Venus; Earth; Mars; Jupiter; Saturn; Uranus; Neptune; Enceladus; Mars: climate change; Venus: Water loss

Figure 4.1: A comparison of the sizes of the Sun (left, only part of it is seen) and the planets. Source: wikimedia.

Arnold Hanslmeier

4.1   Overview of the Solar System

4.1.1   Classes of Objects in the Solar System

Let us consider the different classes of objects in the Solar System (Fig. 4.1):

- The Sun: the central object, host star of the Solar System, contains about 99.8 % of the total mass of the Solar System.

- Planets: there are eight planets orbiting the Sun. Mercury, Venus, Earth, Mars, Jupiter, Saturn, Uranus, Neptune. Planets are spherical, they have removed other small bodies from their orbits.

- Dwarf Planets: they are smaller in diameter than other planets, not perfectly spherical and not removing small bodies from their orbits. Pluto has now been classified as a dwarf planet as well as Ceres, which was classified as asteroid orbiting the Sun between Mars and Jupiter.

- Asteroids or Minor Planets. Only few of them are larger than a few 100 km, their shape ranges from almost spherical to completely aspherical. They are concentrated in several belts like the main asteroid belt between the orbits of Mars and Jupiter or the Kuiper belt outside the orbit of Neptune.

- Satellites: per definition they move about their parent body which can be a planet, dwarf planet and even an asteroid. Some of the largest satellites are larger than the dwarf planets.

- Comets: evaporation of volatile elements from the surface causes their spectacular long tail.

- Meteorites, interplanetary dust: small particles with sizes ranging from mm to few m.

The Sun will be discussed separately.

4.1.2   How to Observe Planets?

The telescopic observation of planets goes back to 1609 when Galileo Galilei first observed Jupiter and its four largest satellites, now known as Galilean Satellites. Before the era of space missions, only telescopic observations provided information about physical parameters of the planets like chemistry of their atmosphere, rotation, temperature *etc.* From position measurements and the study of their motions the masses were derived. The era of observations of extrasolar planets has begun resembling with the era of telescopic observations of planets in our Solar System. We summarize how the different physical parameters can be derived from pure ground based observations:

- Mass: can be obtained from Kepler's third law; M_1 the mass of the Sun and M_2 the mass of a planet. a is the distance of the planet from the Sun (the semi major axis of its orbit to be more precise) and T is the period of revolution.

$$\frac{a^3}{T^2} = \frac{G}{4\pi^2} \left(M_1 + M_2 \right) \tag{4.1}$$

Consider a system Planet-Satellite. The mass of a planet M_1 can also be derived from the motion of a satellite around it. Then M_2 is the mass of a satellite, a the semi major axis of the satellite's orbit and T the period of revolution of the satellite about the planet. Jupiter has the largest mass of all planets in the solar system, about 300 times that of the Earth.

- Temperature: this follows from the IR radiation emitted by a planetary surface. Another way to estimate planetary temperatures is to use the amount of solar energy that it receives being given by S/r^2 where S denotes the solar constant which is the energy we obtain on Earth ($S = 1370\,\mathrm{W/m^2}$) and r is the distance of a planet from the Sun in AU[1]. This energy is reflected back absorbing some amount which heats the surface. The albedo q measures the ratio between energy reflected and absorbed, $q = 1$ means that all energy is reflected, $q = 0$ that all energy is absorbed. Therefore, neglecting possible internal energy sources there is equilibrium:

$$\sigma T^4 = 4\pi R^2 (1 - q) S \tag{4.2}$$

where R is the planetary radius. Let us discuss Eq. 4.2 in more detail. The term on the left hand side (σT^4) is known from Planck's radiation law. It denotes the amount of energy emitted by a body at a given temperature T integrated over all wavelengths. This amount of energy depends on (i) the surface of a planet, given by $4\pi R^2$, (ii) the amount of solar energy received at the planet's surface (S/r^2) and by the amount of energy that is absorbed $(1 - q)$. The real surface temperatures might be quite different from this since planets may have a dense atmosphere (*e.g.* Venus), they rotate (which reduces temperature differences between day and night), and they do not take into account warming by greenhouse gases. Nevertheless, this simple calculation gives an initial estimate being very useful when considering extrasolar planets. From observation we must estimate the albedo q.

- Atmospheric chemistry: can be studied by spectral observations. However, one has to carefully eliminate terrestrial lines.

- Rotation: can be studied by simply analyzing permanent surface features or by observing spectral lines that were taken from opposite parts of the planetary disk in a telescope perpendicular to the planet's equator. The giant gas planets such as Jupiter, Saturn, Uranus and Neptune do not rotate like a solid body but differentially, faster near the equator than near the poles.

Other important parameters like planetary magnetic field can only be measured *in situ* by satellite missions.

Given the parameters of the Earth's orbit (distance to Sun $= 150 \times 10^6$ km, orbital period $= 1$ year), calculate the mass of the Sun from equation 4.1.

Imagine the distance of a planet to the Sun is (i) twice the times that of the Earth, (ii) half the times that of the Earth. What would be the corresponding solar constant S there?

Imagine the Sun's temperature would increase by 10%. How would the solar constant change?

---

[1] 1 AU is the mean distance Earth–Sun=150 000 000 km

Table 4.1: Some important parameters of the planets in the Solar System. D ist the distance from the Sun, P the period of revolution about the Sun, R the planetary radius, and P_{Rot} the rotation period of the planet.

Planet	D $\times 10^6$ km	P	R (km)	P_{Rot}
Mercury	57.91	88.0 d	4879 (0.38)	58.65 d
Venus	108.21	224.7d	12742 (1.0)	-243.02 d
Earth	149.6	365.25 d	12742	23h56m
Mars	227.92	687.0 d	6780 (0.53)	24h37m
Jupiter	778.57	11.75 y	139822 (10.97)	9 h 55m
Saturn	1433.53	29.5 y	116464 (9.14)	10h 40m
Uranus	2872.46	84 y	50724 (3.98)	-17.24 h
Neptune	4495.06	165 y	49248 (3.87)	16.11 h

4.2 Terrestrial Planets

The terrestrial planets are Mercury, Venus, Earth and Mars. They range in size from about 5,000 to 13,000 km. Three of them have an atmosphere (Venus, Earth, and Mars). Their surface is solid and consists of rocks. The basic properties of the terrestrial planets have been elaborated in Table 4.1.

4.2.1 Earth

The Earth is the largest of the terrestrial planets having the mean radius as 6371 km, the mass 5.97×10^{24} kg, and the density $5.5\,\mathrm{g\,cm}^{-3}$. The oceans comprise 2/3 of its surface. Over the whole evolution, the mass of land has increased steadily, during the past two billion years, the total size of the continents has doubled.

Like the other terrestrial planets, the interior of the Earth (see Fig. 4.2) can also be divided into layers:

- the lithosphere is found at a depth between 0-60 km,

- the crust between 5 and 70 km,

- the mantle between 35 and 2890 km,

- the outer core between 2890 and 5100 km and

- the inner core between 5100 and 6371 km (density $\sim 12\,\mathrm{g\,cm}^{-3}$).

Why is the Earth hot in deeper layers? First because it has not cooled out since its formation because of its size and mass. Furthermore, internal heat is produced by radioactive decay of ^{40}K, ^{238}U, ^{232}Th. The rigid crust is very thin compared with the other layers, beneath oceans it is only 5 km, beneath continents ~ 30 km and beneath mountain ranges (Alps) up to 100 km. The atmosphere is composed of 78 % nitrogen and 21 % oxygen.

The Earth is the only known object in the universe where water can be found in liquid state on its surface. If this is a prerequisite for life, then the chances to find life on our neighbor planets Venus and Mars are very low.

We have already discussed in the preceding chapter possible cosmic catastrophes that could lead to mass extinction. It is also well known that the climate on Earth is not stable.

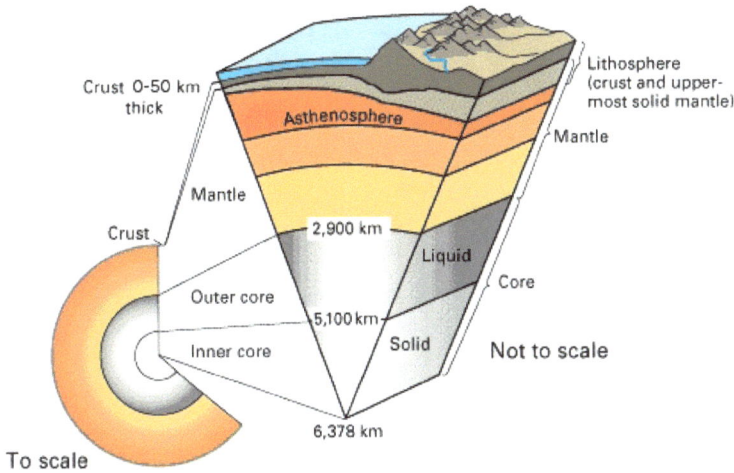

Figure 4.2: The interior of the Earth can be probed using the propagation of seismic waves. Credit: D. Reed.

The origin of ice ages can be explained by a combination of two effects (Fig. 4.3): (i) the varying obliquity of the Earth's axis of rotation. This value can vary by about ±2 degrees around the present day value of 23.5 degrees. The Moon has a stabilizing effect on the obliquity of the Earth's axis. (ii) The eccentricity of the Earth's orbit about the Sun also changes. Higher eccentricity means larger differences between perihelion (when the Earth being nearest to the Sun) and aphelion (when the Earth is farthest to the Sun).

4.2.2 Mercury

Mercury is the nearest planet to the Sun and the temperature on the dayside is up to 450° C and on the nightside –180° C. This extremely high temperature contrast is a consequence of lacking an atmosphere. In 1974 and 1975, the US-Mariner 10 spacecraft made three passes by Mercury, sending back photos of 45 percent of the planet's surface.

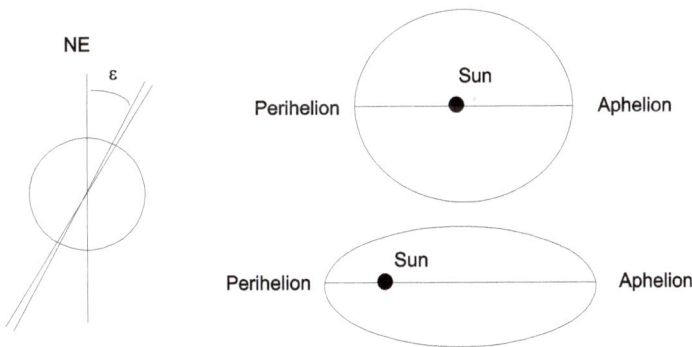

Figure 4.3: Left: change of the Earth's rotational axis tilt; ϵ is the angle between the axis of rotation and the normal to the ecliptic (NE ecliptic north pole), right: effect of the eccentricity e on the perihelion and aphelion distance..

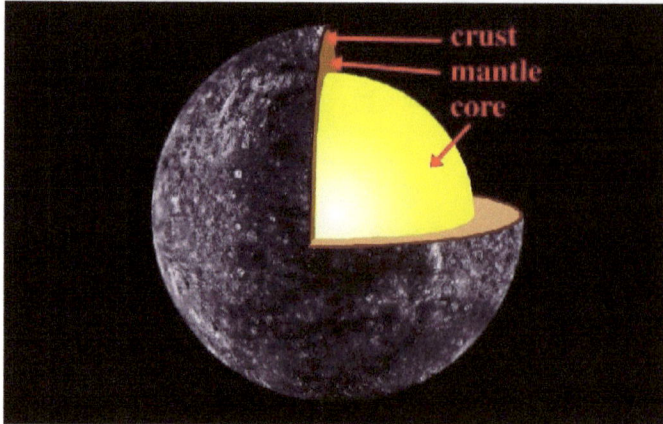

Figure 4.4: The interior structure of planet mercury showing a huge nucleus. Credit: NASA.

In 1991, planetary scientists Duane Muhleman and Bryan Butler from Caltech and Martin Slade from the Jet Propulsion Laboratory, investigated Mercury using a radar system consisting of a 70-meter (230-foot) dish antenna at Goldstone, CA. The beam of 8.5-GHz microwaves sent from Goldstone bounced off Mercury and was collected at the VLA[2] to produce a radar image of the planet. They found some strong radar reflecting regions on the planet's surface. In images, the bright red dot at the top of the image indicates strong radar reflection at Mercury's north pole. This reflection resembles the strong radar echo seen from the ice-rich polar caps of Mars.

This led to the assumptions that ice could persist inside deep craters near the poles of Mercury. Normal ice absorbs radar signals, but ice at extremely low temperatures reflects them.

Radar observations of Mercury have revealed the presence of anomalous radar reflectivity and polarization features near its north and south poles. It can be expected that certain regions near the poles of Mercury never get warmer than 170 K. Therefore it is argued that maximum surface temperatures in shaded cratered regions near the poles below 110 K and at temperature lower than 112 K water ice could be stable to evaporation for several billion years.

Exogenic sources of water for Mercury's polar ice have been discussed. Continual micrometeoritic bombardment of Mercury over the last 3.5 billion years could have resulted in the delivery of $(3-60) \times 10^{16}$ g of water ice to the permanently shaded regions at Mercury's poles (equivalent to an average ice thickness of 0.8-20 m).

The MESSENGER probe (Mercury Surface, Space Environment, Geochemistry and Ranging) which was launched 2004 became in 2011 the first spacecraft to orbit Mercury. The various colors seen across Mercury are due to different mineral compositions of the geologic regions, younger surface materials being brighter at visible wavelengths and less affected by space weathering. Mercury has an exceptionally large iron core, a solid layer of iron sulfide and a thin silicate crust. Many craters seem to have tilted over time, so Mercury is much more active than the Moon. Bright patches appear inside Mercury's shady craters, a potential indicator that water ice is present.

---

[2]The Very Large Array

Figure 4.5: Comparison between Venus (radar image) and Earth. Credit: NASA

The interior structure of Mercury is shown in Fig. 4.4.

4.2.3   Venus

Venus is often referred to as "sister Earth" because it is similar to earth in size (Fig. 4.5). In former times, it was assumed that Venus contains much water, in the shapes of oceans and swamps. A hot and wet climate was expected on that planet. Unfortunately, because of its dense clouds, it was impossible to observe any details on its surface - but this fact inspired even people more. However, observations made from ground by radar and from satellite missions and even successful landings revealed a completely different picture of that planet. Its surface being extremely dry and resembles the surface of Mercury and the Moon. Being also extremely hot, the conditions for life on the surface are not very bleak. The surface pressure of the atmosphere is 90 times that of the Earth's[3], and, because of the dense atmosphere consisting mainly the greenhouse gas CO_2, the surface temperature is about 460° C. It is interesting to note, that this high surface temperature can be found on any place on the surface of Venus, there are no big differences between *e.g.* the equator and the poles. Therefore, contrary to Mercury, it is unlikely to find ice or water near the poles of Venus. Venus rotates in 243 Earth days in the direction opposite to its orbital motion and one revolution around the Sun lasts 224.7 days.

Let us compare the water content in the atmosphere of Venus with that in the terrestrial atmosphere:

- Earth: average surface temperature 288 K, water vapor content in the troposphere up to 4% (Lodders and Fegley, 1998 [66]).

- Venus: surface temperature range from 660 K (on the high Maxwell mountains) to 740 K (in the plains), water content in the subcloud atmosphere about 30 ppmv[4].

---

[3]That corresponds to the pressure at a water depth of 900 m on Earth.
[4]parts per million by volume

Figure 4.6: Mars observed with the Hubble Space telescope (HST). The polar caps are clearly seen as well some dust in the atmosphere and surface features. Credit: NASA, HST.

4.2.4 Mars

Mars is the second closest in similarity to Earth. Due to the turbulent Earth's atmosphere, it is difficult to observe details about its surface from earthbound observatories. Moreover, because of its elliptic orbit, the closest distance to Earth can range from about 56 Million to more than 100 Million km. Turbulence in the Earth's atmosphere blurs and distorts the image (this effect is known as seeing in astronomy) and can lead to completely wrong interpretations of observations. A famous example of such wrong interpretation is the canali on Mars. Schiaparelli observed Mars during its approach to Earth in 1877. He claimed to have detected on the surface of Mars a network of channels which he called canali. Later, Lowell and Flammarion suggested that these canali were constructed by intelligent Martians to distribute water on the dry planet. From that time on (beginning of the twentieth century) the legend of little green men on Mars was created.

Many attempts have been made to investigate the red planet through satellite missions. There seems to exist a special Mars Curse: until 2006 only 18 out of 37 launch attempts to reach Mars have been successful. The first images of the surface of Mars from a satellite mission were obtained in 1965 by Mariner 4 (US). In 1971 for the first time a satellite could be brought into orbit around a planet (with exception of Earth of course). The results were disappointing because the surface of Mars appeared more like that of the Moon and there were no signs of the canali claimed to have been observed (see also HST image in Fig. 4.6).

On a nice summer day, the temperature on Mars may rise up to 0° C, however, during a Martian night it may reach $-100°$ C. There are river like channels on its surface hinting that this planet has undergone large climatic variations in the past with episodes of liquid water on its surface[5]. Today, because of the low atmospheric pressure (only 1% of the pressure on the surface of the Earth) water cannot exist in liquid state on Mars. For liquid water on the martian surface, the atmosphere must become much denser than it is today. Water can only sublimate on the surface of Mars *i.e.* it makes a phase transition from solid (ice) to

---

[5]These channels have nothing in common what has been claimed by the optical illusion of Martian channels observed from Earth

gas (vapor) or from gas to solid.

Mars has two polar caps. The permanent portion of the north polar cap consists almost entirely of water ice. In the northern hemisphere winter, an additional coating of frozen carbon dioxide about one meter thick is gained. The south polar cap also acquires a thin frozen carbon dioxide coating in the southern hemisphere winter. Beneath this is the perennial south polar cap, which is in two layers. The top layer consists of frozen carbon dioxide and is about 8 meters thick. The bottom layer is very much deeper being made of water ice. In 2005 data from NASA's Mars Global Surveyor and Odyssey missions revealed that the carbon dioxide ice caps near Mars's south pole had been melting for three summers in a row. This can be regarded as a sign of long-term increase in solar irradiance that affects both the Earth and Mars.

4.2.5   Dry Venus - Humid Earth - Climate Changes on Mars

After discussing the terrestrial planets let us address to the question why they appear so different concerning their water content. One important criterion is of course the surface temperature. On Mercury, the temperatures are so high because of its closeness to the Sun, that water cannot exist besides, possibly, near the poles.

It is reasonable to assume that Venus and Earth and even Mars accumulated similar amounts of primordial water during their accretion phase. So why did Venus lose its water?

The distance of Venus from the Sun is about 2/3 the Earth's distance from the Sun. Theoretically, Venus could have quite favorable conditions for liquid water to exist on its surface. But there are two effects need to be considered, first, the present dense atmosphere of Venus with the greenhouse gas carbon dioxide causing too high surface temperatures. Second, this unfavorable condition might has been different in the early evolution of Venus. The luminosity of the early Sun was only about 70% of its present value. Therefore, on early Venus, water oceans could have existed. But the Sun became more luminous and due to its closer distance, Venus received more radiation from the Sun than Earth. Especially, the UV part of radiation has to be mentioned here. It causes a splitting of the water molecules which led to a runaway greenhouse effect. The lighter hydrogen escaped from the atmosphere, the heavier oxygen was bound with surface minerals. The other question concerns the evolution of Venus's atmosphere. This has to be seen in connection with plate tectonics. On Earth, gases like carbon dioxide are released but are also absorbed by its active plate tectonics. On Venus, there have not been any signs of plate tectonics in recent time. Therefore, its atmosphere became enriched with that greenhouse gas.

The problem of a wet early Venus was studied by several authors. Pioneer Venus mass spectrometer data suggest that the ratio of deuterium to hydrogen, D/H ration is 1.9×10^{-2}. This is about 120 ± 40 times the value found for Earth. A fractionation process during evaporation of diffusion causes the lighter hydrogen preferentially to escape leaving the heavier deuterium. Venus could once have had the equivalent of an at least 4 m thick ocean (the maximum depth however could have been more than 500 m). It has to be stressed that the D/H value alone could be problematic in order to estimate the amount of water because there are examples of interplanetary dust particles and comets with high D/H ratios.

A study of the surface mineralogy of Venus could provide another proof for the existence of large amounts of water in its early phase.

Another important isotope for studying the water-history of Venus is ^{40}Ar. The atmospheric abundance of ^{40}Ar is only $\sim 25\%$ of the radiogenic gas produced inside Venus. This may imply that Venus is not thoroughly degassed as its interior has not been dried over time. A dry interior of Venus can be explained by a near head-on collision of two large planetary embryos. Such a collision would be sufficiently large to melt totally and briefly vaporize a major proportion of both the bodies.

While Earth and Mars (due to the SNC meteorites been ejected from Mars and found on Earth) have been studied extensively by comparative planetology, the most earthlike planet Venus has still been poorly understood. A recent summary was given by [20]. Observations of the dayside of Venus performed by the high spectral resolution channel (-H) of the Visible and Infrared Thermal Imaging Spectrometer (VIRTIS) on board the ESA Venus Express mission[6] have been used to measure the altitude of the cloud tops and the water vapor abundance around this level with a spatial resolution ranging from 100 to 10 km. CO_2 and H_2O bands between 2.48 and 2.60 μm have been analyzed to determine the cloud top altitude and water vapor abundance near this level. The following results have been found ([27]):

- At low latitudes (40^0) mean water vapor abundance is equal to 3 ± 1 ppm and the corresponding cloud top altitude at 2.5 μm is equal to 69.5 ± 2 km.

- Poleward from middle latitudes the cloud top altitude gradually decreases down to 64 km, while the average H_2O abundance reaches its maximum of 5 ppm at 80^0.

Mars

The story whether there is water on Mars or not, and if yes, how much, is not completely finished yet. In 1976 the first unmanned spacecrafts landed on the Martian surface (Viking 1 and Viking 2). Mars appeared as a dry, rocky desert with no signs of water. In the 1990s however, that picture changed again as many features on the Martian surface were detected that could have been only formed by water (Fig. 4.7). In 2006, the NASA Mars Global Surveyor orbiter, found evidence of water flowing fleetingly on the surface. Comparing images of the side of a crater taken in 2001 and 2005 the later showed gullies apparently caused by water bursting out of the crater wall. These gullies can be seen in Fig. 4.8.

Mariner 9 images showed equatorial sinuous channels on Mars. There may be two possible stable climates: one that resembles the present day climate on Mars, the other which an atmospheric pressure of about 1 bar. The triggers for a transition from one state to another are the changes noticed in the obliquity, solar luminosity and albedo variations of the polar caps and also periods of extensive volcanism.

Generally, the obliquity is a primary factor for climate changes. Obliquity changes result into drastic climate changes because of a runaway sublimation of permanent CO_2 ice. Simulation lead to ring-like structures of CO_2 ice at mid latitudes. In Fig. 4.9 the distribution of water over the whole planet is shown. Up to 16% of water is found near the poles, however there is also a large area near the equator which has concentrations up to 10%. A possible explanation for that is that about one million years ago, the rotational axis of Mars was tilted by 35^0 and this caused the water near the poles to melt and distribute to lower latitudes. These data were obtained with a neutron spectrometer. When cosmic rays collide into the nuclei of atoms on a planet's surface some are ejected with enough energy to reach the orbiter (Odyssey spacecraft[7]). Elements create their own unique distribution of neutron energy - fast, thermal or epithermal - and these neutron flux signatures are shaped by the elements that make up the soil and configure their distribution. Thermal neutrons are low-energy neutrons in thermal contact with the soil. Epithermal neutrons are intermediate,

---

[6]Was launched on Nov. 5, 2005
[7]Launch: April 7, 2001; Arrival: October 24, 2001

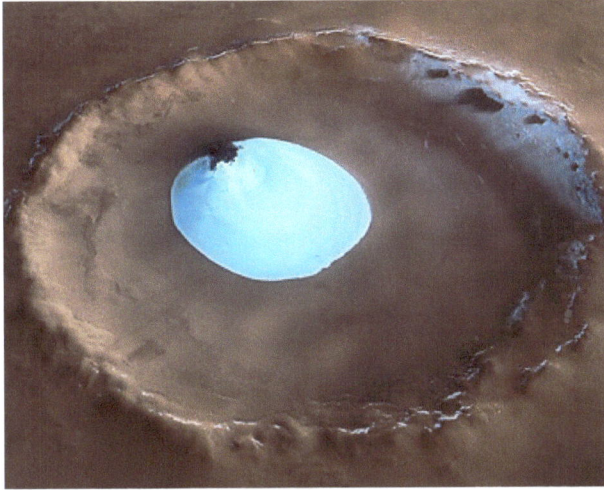

Figure 4.7: Water ice in crater at Martian north pole. The crater is 35 km wide having a maximum depth of approximately 2 km beneath the crater rim. The circular patch of bright material located at the center of the crater is residual water ice. The colors are very close to natural, but the vertical relief has been exaggerated thrice. The view is looking east. Credits: ESA/DLR/FU Berlin (G. Neukum)

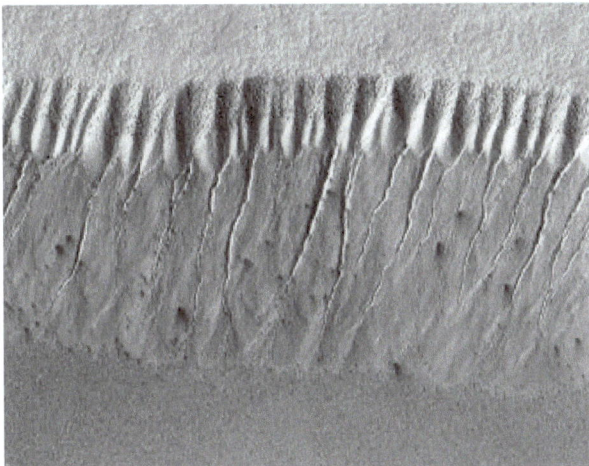

Figure 4.8: These gully landforms near the south polar region of Mars show evidence of geologically recent seepage and runoff of liquid water. Today, however, water on Mars appears to exist mainly as subsurface ice.

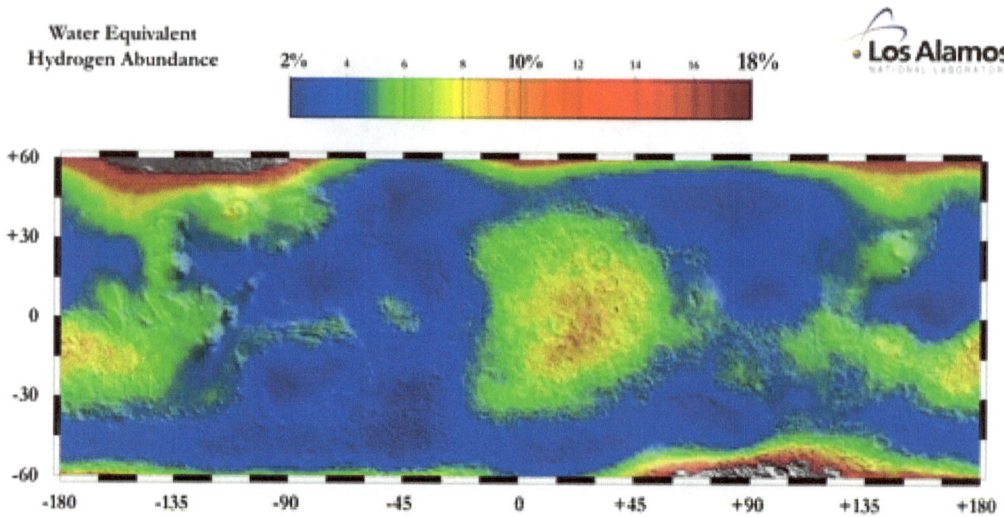

Figure 4.9: Distribution of water on Mars: overlay of water equivalent hydrogen abundances and a shade relief map derived from MOLA (Mars Orbital Laser Altimeter) topography. Mass percents of water were determined from epithermal neutron counting rates using the Neutron Spectrometer aboard Mars Odyssey between February 2002 and April 2003.

scattering down in energy after bouncing off soil material. Fast neutrons are the highest-energy neutrons produced in the interaction between high-energy galactic cosmic rays and the soil.

By looking for a decrease in epithermal neutron flux, hydrogen can be located because hydrogen in the soil efficiently absorbs the energy from neutrons, reducing their flux in the surface and also the flux that escapes the surface to space where it is detected by the spectrometer. Since hydrogen can easily be found in the form of water-ice at high latitudes, the spectrometer can measure directly, about one meter or so deep into the Martian surface, the amount of ice and the changes in brought by the seasons. The principle of Gamma Ray Spectroscopy is shown in Fig. 4.10.

There seems to be a lot of ground ice in the martian soil. Observational features such as viscous creep support that hypothesis. Such a reservoir would be equivalent to a global layer of water 10-40 cm, distributed over the whole martian surface. This reservoir was created during the last phase of large high orbital obliquity less than 100 000 years ago and now it it is on the verge of decay. Let us give a comparison of these numbers with numbers for the Earth. If all ice on Earth melts, the level of the oceans would rise by 64 m.

At present, Mars does not have a global magnetic field. However, in its early phase a dynamo mechanism could have worked causing a global martian magnetic field. Such a field might have effectively shielded the planet's atmosphere from solar wind particles. Later in the history of martian evolution, cooling and/or solidification of the core would cause dynamo extinction and expose the Martian atmosphere to the full solar and galactic particle flux. This might have had considerable influence on the global martian climate.

A new analysis of the magnetic field signature over appropriately sized impact basins and volcanoes on Mars prove that the dynamo lasted until 3.77 Ga, which may explain why

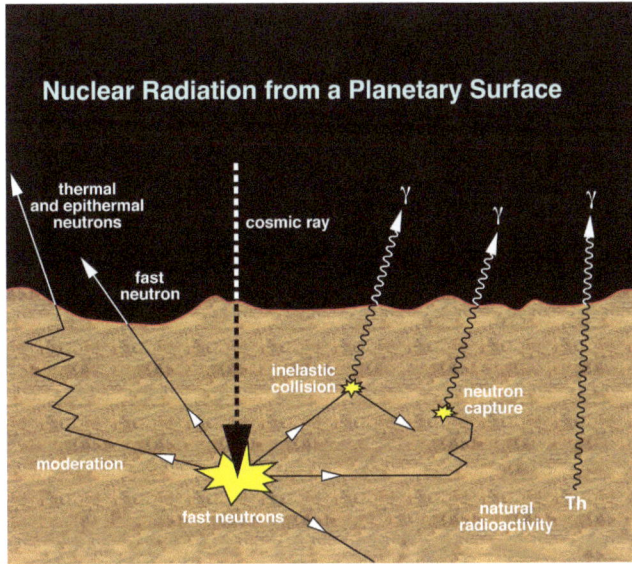

Figure 4.10: Principle of Gamma Ray Spectroscopy to probe the subsurface composition of planets. Credit: NASA.

most of the liquid water surface activity persisted through that epoch [59]

Compare the UV flux received at the orbits of Venus, Earth and Mars!

Clouds and Photolysis

In the classical work of Kasting *et al.*, [47] the extension of the habitable zone in the solar system was discussed. The authors assumed a CO_2, H_2O, N_2 atmosphere on a hypothetical earthsized planet and studied the effect of different distances from the Sun on the planet's surface temperature. The different solar flux has two effects:

- If the planet is too close to the Sun: Photolysis of water, evaporation of oceans. That happened to Venus.

- If the planet is too far away, then CO_2 cloud start being formed increasing strongly the albedo of a planet and cooling it. That happened to Mars.

Therefore habitable planets can exist in a zone between 0.95 and 1.37 AU from the Sun. This idea has also strong consequences for the Earth:

- If the distance of Sun-Earth had been 4-7% closer oceans would have never condensed on Earth.

- If the Sun-Earth distance had been only 1 to 2% farther, Earth would have become a global snowball.

Figure 4.11: Comparison between the sizes of giant planets: Jupiter, Saturn and Uranus. The size of the Earth would be about 1/4 the size of Uranus.

4.3   Giant Planets

Giant planets having no solid surfaces were able to capture a larger amount of gases and volatile materials from the nebula from which the Solar System is formed. There exist two groups of giant planets in the Solar System:

- Gas giants: Jupiter and Saturn; both are similar in size composed primarily of hydrogen and helium.

- Ice giants: Uranus and Neptune; both are similar in size and contain larger amounts of water and other ices.

A comparison of sizes of the giant planets is given in Fig. 4.11 and data can be found in Table 4.1.

4.3.1   Jupiter and Saturn

Jupiter is the largest of the eight planets and its diameter is about one tenth that of the Sun or about ten times that of the Earth. Its chemical composition is quite similar to that of the Sun. Jupiter consists mainly of H and He, only 2 percent of its mass is made up of other elements. Viewed through a telescope, parallel bands can bee seen, the darker bands are called belts, the lighter ones zones. At the edges or within belts many small clouds appear. In the southern hemisphere the Great Red Spot (Fig. 4.14) is observed which has an extension of $25,000 \times 12,000$ km. It is being seen since nearly three centuries. The great Red Spot is a giant anticyclone, circulating like a vortex.

Saturn is very well known because of its magnificent ring systems, but it was detected later that all giant planets are surrounded by equatorial ring systems. Cloudlike features in its atmosphere appear less prominent than in Jupiter's atmosphere. Jet streams were discovered in its atmosphere being similar to jet streams in the terrestrial atmosphere. These jet streams can be easily explained by an alternate pattern of high and low pressure systems.

The Galileo probe descended through Jupiter's atmosphere. It was launched on October 1989 by the Space Shuttle Atlantis and arrived at Jupiter on December 7, 1995 assited gravitational flybys of Venus and Earth. This was the first direct exploration of the deeper layers of Jupiter's atmosphere. As expected, an increase in the outside pressure and temperature was recorded on decreasing the altitude of spacecraft. Dense layers of clouds were found tbeing separated by a relatively clear atmosphere. These cloud layers are composed of different chemical substances. Why different cloud layers in Jupiter's atmosphere exist? The answer is surprisingly simple. In the Earth's atmosphere, water is the only substance that can condense into clouds, however, in Jupiter's atmosphere there are many species of volatiles that can condense. Each kind of volatiles condenses at a particular temperature and pressure. The basic composition of Jupiter's atmosphere is structured in different layers (see also Fig. 4.12):

- Stratosphere: from about 0 to infinity.

- at h=0 km there is haze.

- Troposphere: reaches up to h=0 km and consists of different cloud layers

    - Ammonia ice -40 to -20 km
    - Haze: at about -50 km; here the pressure corresponds to atmospheric pressure at sea level on Earth.
    - Ammonium hydrosulfide ice layer: - 60 to -50 km
    - Water ice layer: -70 to -60 km; here the pressure is about five times the atmospheric pressure at sea level on Earth.
    - Water/ammonia droplets: occur between -90 and -80 km. The pressure is about 7 to 8 atm.
    - Below 100 km: gaseous hydrogen, helium, methane, ammonia and water

The distance of a planet from the Sun helps in defining its tropospheric temperatures. This temperature determines the altitude at which substances like water or ammonia condense to form a cloud layer. Methane exists in gaseous form only throughout the warmer tropospheres of Jupiter and Saturn.

Looking at pictures of Jupiter and Saturn (Fig. 4.16) we find colorful clouds, whereas ices like water and ammonia appear to be white. The colors are caused by impurities in the ice crystals. These impurities consist of sulfur, phosphorus and organic materials produced by photochemical action of sunlight on atmospheric hydrocarbons. Solar UV radiation is absorbed by hydrocarbons such as methane, acetylene, ethane *etc*. These molecules are destroyed starting to form complex organic compounds that condense into solid particles. Such types of reactions are well known for the terrestrial atmosphere known as photochemical smog. The chemicals of photochemical smog on Earth include nitrogen oxides, volatile organic compounds (VOCs), ozone and peroxyactyl nitrate. Cars produce nitrogen oxides, VOCs are emitted from paint, gasoline and pesticides. Examples of VOCs being quite hazardous are: benzen, polycyclic aromatic hydrocarbons (PAHs). Benzene increases the chances to get leukemia, while PAHs cause cancer.

Some reactions which are of major importance in the upper atmosphere of Jupiter are:

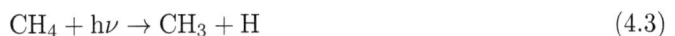

$$CH_4 + h\nu \rightarrow CH_3 + H \qquad (4.3)$$

then for example the reaction

$$CH_3 + CH_3 + M \rightarrow C_2H_6 + M \tag{4.4}$$

or

$$CH_3 + H + M \rightarrow CH_4 + M \tag{4.5}$$

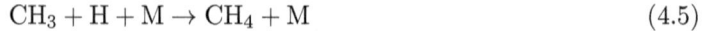

occur. Another example of photolysis with subsequent reactions is

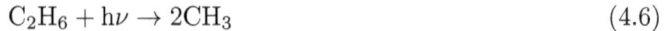

$$C_2H_6 + h\nu \rightarrow 2CH_3 \tag{4.6}$$

followed by several reactions.

Giant planets rotate rapidly. The Coriolis force is a deflection of moving objects when they are viewed in a rotating reference frame. The acceleration of a particle in the rotating system is given by

$$a_c = -2\,\mathbf{\Omega} \times \mathbf{v} \tag{4.7}$$

$\mathbf{v}$ is the velocity of the particle, $\mathbf{\Omega}$ is the angular velocity vector. The magnitude of $\mathbf{\Omega}$ is equal to the rotation rate ω. The cross product can be evaluated as the determinant of a matrix:

$$\mathbf{\Omega} \times \mathbf{v} = \begin{pmatrix} \mathbf{i} & \mathbf{j} & \mathbf{k} \\ \Omega_x & \Omega_y & \Omega_z \\ v_x & v_y & v_z \end{pmatrix} = \begin{pmatrix} \Omega_y v_z - \Omega_z v_y \\ \Omega_z v_x - \Omega_x v_z \\ \Omega_x v_y - \Omega_y v_x \end{pmatrix} \tag{4.8}$$

where $\mathbf{i}$, $\mathbf{j}$, $\mathbf{k}$ are unit vectors in the x,y and z directions.

The Coriolis effects in the atmospheres of giant planets cause strong zonal winds. On Jupiter and Saturn the winds are equatorial westerlies, on Jupiter up to 550 km/, and on Saturn up to 1,700 km/h.

How can we calculate wind speeds on distant planets? We just need to track individual clouds. By measuring the position of these clouds the amount of their movement during an interval (*e.g.* a day) is noted. This has to be subtracted from the rotation of the planet. The rotation of a planet may be determined by bursts of radio waves caused by the rotation of a planet's magnetic field.

The atmosphere of Jupiter can be probed by radio occultation experiments using the Voyager spacecrafts[8] [63]. This method is often used even for the determination of the upper Earth's atmospheric composition. In Fig. 4.13 this method has been demonstrated. Passing through an atmosphere an electromagnetic wave will be diffracted and phase shifted.

Both Jupiter and Saturn have a large source of internal energy (also Neptune has one). The equilibrium temperature of Jupiter is 109 K, but an average temperature of 124 K is measured. This 15 K may not seem much, but let us compare the radiated energies. The radiated energy is given by Stefan's law:

---

[8]The two spacecraft were launched in August and September 1977

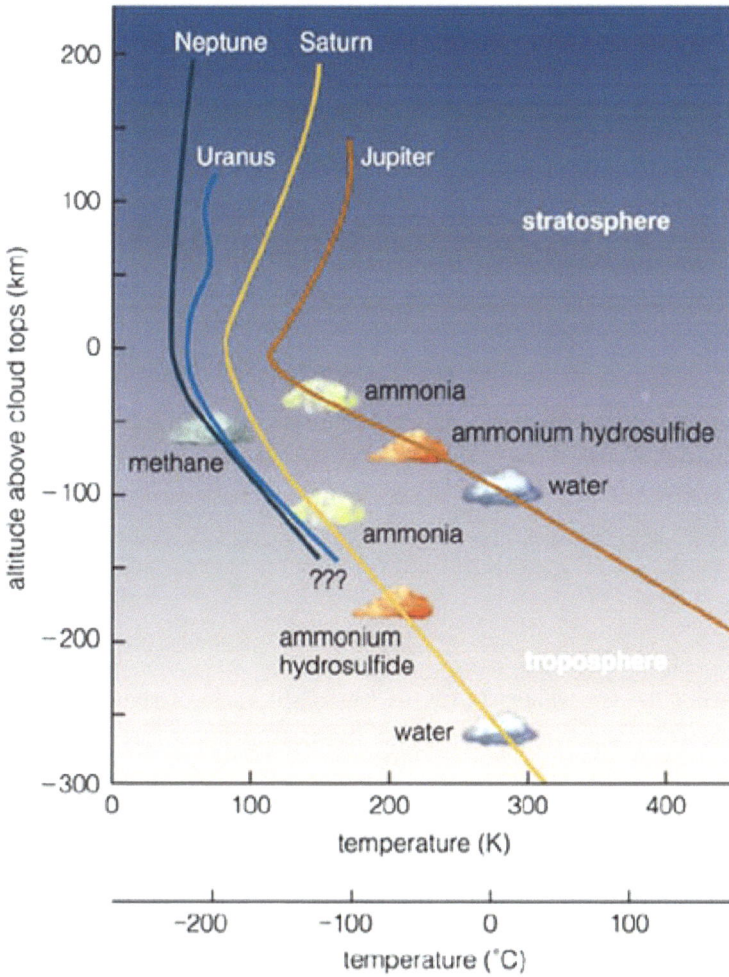

Figure 4.12: The atmospheres of the giant planets. Credit: astronomy online

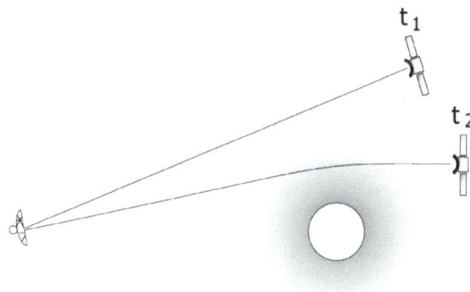

Figure 4.13: The radio occultation method. At position 2 the radio beam is influenced by the atmosphere of a planet permitting to measure physical parameters in the atmosphere.

Figure 4.14: Jupiter with red spot. The spot is a swirling cold storm system with higher temperatures at its core, new thermal imaging studies have shown. Credit NASA Voyager

$$E = \sigma T^4 \tag{4.9}$$

Now, by comparing the two values 124 K to 109 K the ratio of the two different energies becomes:

$$\left(\frac{124}{109}\right)^4 = 1.67 \tag{4.10}$$

Jupiter is radiating two-thirds more energy into space than it absorbs from the solar irradiation. So the weather on Jupiter is powered by almost half from internal energy. For Saturn and Neptune this value is even higher, about 2. The source for this internal energy is the conversion of gravitational potential energy into thermal energy.

The interiors of the giant planets are hot and dense. The interior of Jupiter (Fig. 4.15) and Saturn consists of molecular hydrogen. In deeper layers hydrogen becomes metallic. Jupiter's core consists of liquid water and rock at a temperature of about 20,000 K. Here, the pressure is several tens of megabars. A pressure of 1 bar corresponds about the atmospheric pressure at Earth's sea level. At a depth of 10,000 in the ocean the pressure is 1,000 bars. Water may still be liquid at high temperatures if the pressure is high enough[9]. The pressure at Jupiter's center is about 70 megabars while the temperature 20,000 K. At a pressure of 4 megabars and temperature 10,000 K hydrogen molecules loose their electron and hydrogen becomes electrically conducting like a liquid metal. This occurs at a depth of about 20,000 km in Jupiter's atmosphere and 30,000 in Saturn's.

Helium can be also compressed to a liquid state but does not reach metallic state under the physical conditions existing in the interiors of the giant planets. In Jupiter, because of the higher temperature, liquid helium is mostly dissolved with liquid hydrogen. In Saturn, because of the lower temperature, the helium is less soluble[10]. Therefore, helium can exist in droplets in Saturn, these droplets sink and providing an additional source of internal energy.

---

[9]Like a super pressure cooker

[10]Analogon: you can easily dissolve large amounts of sugar in hot water water relatively little when the water is cold

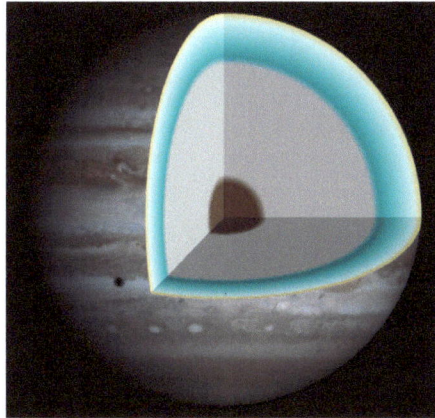

Figure 4.15: The interior of Jupiter. This cut-away illustrates a model of Jupiter's interior. In the upper layers the atmosphere transitions to a liquid state above a thick layer of metallic hydrogen. In the center there may be a solid core of heavier elements. Credit: NASA/R.J. Hall

Jupiter's magnetic field is 20,000 times as strong as Earth's, Saturn's field 500 times. At the cloud tops, Jupiter's field is still 15 times that of the Earth's field on the surface. The magnetosphere of Jupiter extends about 100 times Jupiter's radius. The pressure of the solar wind pushes and compresses a magnetosphere, and planetary magnetic fields also divert solar wind particles which then flow around magnetospheres. We can compare these processes with the way a stream flows around boulders. A rock in a river creates a wake extending downstream. Also a planetary magnetosphere creates a wake. The wake of Jupiter's magnetosphere extends past the orbit of Saturn. Rapidly moving electrons spiral around the direction of magnetic field lines and causing synchrotron radiation. This radiation can be observed as radio waves. In the radio wavelengths, Jupiter would appear twice as large as the Sun on our sky.

There is another interesting effect caused by magnetospheres. Charged particles are trapped in planetary magnetospheres and become concentrated in defined regions, the radiation belts. When in 1973, the Pioneer 11 spacecraft passed trough Jupiter's radiation belts a dose of 400,000 rad was measured which corresponds to 1,000 times the lethal dose for humans. Io is the innermost satellite of Jupiter. Because of extreme tidal forces, Io is volcanic and emits sodium. This element is distributed in a torus around its orbit.

What is the difference between thermal and synchrotron radiation?

4.3.2   Uranus and Neptune

Observations of structure in the atmospheres of Uranus (Fig. 4.17) and Neptune are extremely difficult because of two reasons: they appear as small disks with a pale bluish or green color having no prominent features as was found by space missions. Infrared observations made with the Hubble Space Telescope revealed structures like hazes, bands and small clouds.

The atmospheres of Uranus and Neptune are quite different from that of Jupiter and

Figure 4.16: Saturn. On one of the poles aurorae can be seen. Credit: NASA, ESA, J. Clarke (Boston University), and Z. Levay (STScI).

Saturn. Only few thin clouds have been found to be visible on Uranus, one can be seen into a clear bottomless atmosphere. In greater depths there must be clouds but they can not be seen because of molecular scattering in the clear part of the atmosphere. In the atmosphere of Neptune few high clouds are often observable. Neptune had a Dark Black Spot which has disappeared.

All four giant planets have a core consisting of 10 to 15 times the mass of Earth.

Seasons depend on the tilt of the rotation axis, since the tilt is only 3^0 for Jupiter, no seasonal effects occur there, but are similar to Earth for Saturn and Neptune. Uranus spins on its axis being nearly in the plane of its orbit. The obliquity of its axis is 98^0. It faces 42 years continuous sunshine and 42 years continuous darkness on its polar regions. The poles are warmed up more than the equator. The strange obliquity of the rotation axis of Uranus may have been caused by an impact of a huge planetesimal near the end of its accretion phase.

The atmospheres of Uranus and Neptune appear structured as follows:

- Stratosphere,

- from 0 to 30 km methane ice,

- from -50 to 0 km hydrogen sulfide ice,

- from -120 to -60 km there exists water ice, ammonium hydrosulfide ice,

- at greater depths water, ammonia, hydrogen sulfide droplets

The upper tropospheres are relatively clear, where methane gas is more abundant than in the atmospheres of Jupiter and Saturn. In Neptune, strong easterly winds occur with speeds up to 1,600 km/h.

Neptune has an internal energy source (Fig. 4.18), however Uranus not. Uranus and Neptune consist mainly of water and low density ices. The total amount of hydrogen and He is not more than 1 to 2 times the mass of Earth. The water is in the form of deep

Figure 4.17: Uranus photographed by the HST. Credit: NASA, HST.

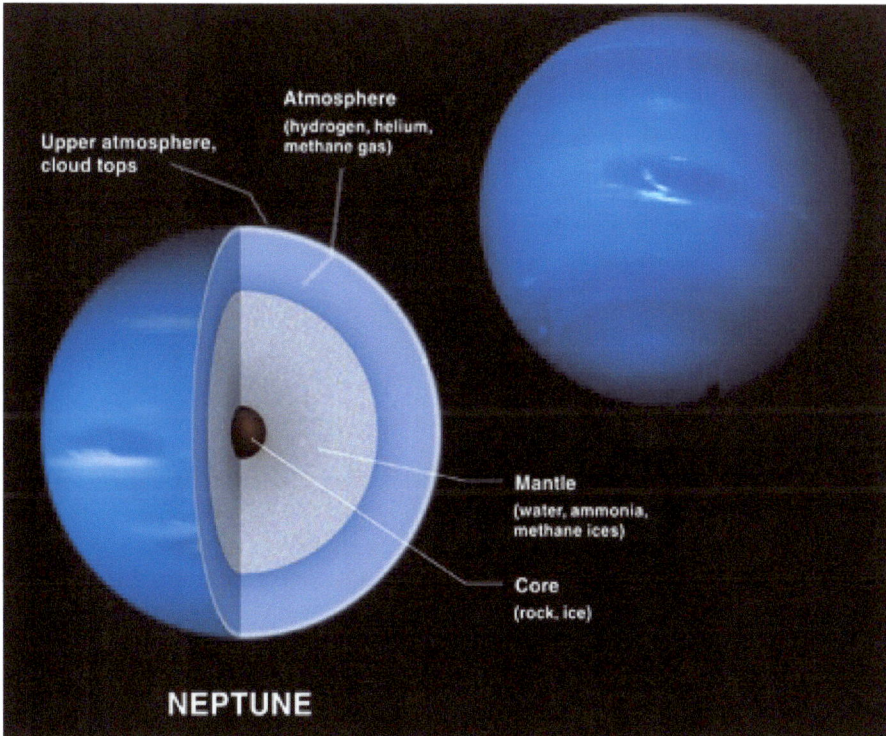

Figure 4.18: Neptune and its internal structure. Credit: NASA.

oceans. Because of the presence of dissolved gases and salts, the oceans are electrically conducting. This saltwater causes the magnetic fields of these planets, whereas metallic hydrogen causes the magnetic fields of the gas giants Jupiter and Saturn. There is another remarkable peculiarity. Neptune's rotation axis is similar to that of Earth, Mars and Saturn. Its magnetic field axis is inclined by 47 degrees.

4.4   Satellites and Rings of Planets

In 1655 Ch. Huygens could clearly see that Saturn is surrounded by a ring. Later D. Cassini found a gap in the ring of Saturn and in 1850 a fainter inner ring was found (Fig. 4.16). Satellites of planets were first reported in 1609 by G. Galilei who found the four large satellites of Jupiter.

4.4.1   Rings

It was suspected that also the other giant gas and ice planets may have rings but until 1977 all attempts to detect them proved in vain. At that time a stellar occultation caused by Uranus was observed and a short time before and after the planet's disk occulted the star, a sharp decrease of the brightness of the star was noticed. This occultation was caused by the rings. Several rings of Uranus were detected in 1986 by Voyager 2 followed by the Hubble Space Telescope as 13 rings of Uranus were found totally. By stellar occultation several properties of rings can be measured such as the duration of the occultation hinting for the width of the ring and the decrease in brightness of a star depending on the ring material. In 1979 a ring around Jupiter was found from Pioneer 11 observations. In 1989, Voyager 2 reached Neptune discovering a ring arc like structure around Neptune. Neptune's rings are faint with the exception of this structure.

But Saturn is the planet having the most spectacular rings. Theses rings are not homogeneous, since the A and C rings containing only hundreds, while the B ring containing thousands of individual ringlets, some of them a few kilometers wide. After about every 15 years, the rings seem to vanish when viewed from Earth since the plane of Saturn's rings lines up with the plane of Earth's orbit. The rings of Saturn are wide, 62,000 km from the inner edge of the C Ring to the outer edge of the A Ring, however, being extremely thin, certainly less than 100 m.

Jupiter's rings are mostly made up of fine dust but not of rocks. Uranus contains 13 rings, most of them being quite narrow consisting of dark material. Similar to Saturn's D Ring, material in the 11th ring may be spiraling into the top of the Uranus atmosphere.

Three rings of Neptune are narrow (few tens of kilometers).

The origin of the rings can be explained as remains of icy moons disrupted by tides. Saturn's rings are found inside the Roche Limit[11]. Considering a satellite orbiting a large planet, we will come to know that the parts of the satellite that are closer to the planet are attracted by stronger gravity from the planet, whereas parts further away are repelled by stronger centrifugal force from the satellite's orbit. Of course satellites (natural as well as artificial) can also orbit a planet inside the Roche limit when these objects are held together by internal forces (cohesion). Examples of such natural satellites are Jupiter's moon Metis or Saturn's moon Pan. Also comets could disintegrate when passing the Roche limit (this happened for example in the case of Shoemaker Levy, Fig. 4.19). Within the Roche limit, no large satellite is able to exist. Almost all planetary rings are located within their Roche limit exception are Saturn's E Ring and Phoebe ring. They could be remnants from the

---

[11] At this limit no stable bodies can exist, because of strong tidal effects

Figure 4.19: A NASA Hubble Space Telescope (HST) image of comet Shoemaker-Levy 9, taken on May 17, 1994, with the Wide Field Planetary Camera 2 (WFPC2) in wide field mode. When the comet was observed, its train of 21 icy fragments stretched across 1.1 million km of space, or 3 times the distance between Earth and the Moon. Source: NASA, ESA, and H. Weaver and E. Smith (STScI).

proto-planetary accretion disc where the planet was formed or the remnants of a broken moon that passed the Roche limit.

Let us consider a mass u with radius r that is attracted by a satellite with mass m at a distance d:

$$F_G = \frac{Gmu}{d^2} \tag{4.11}$$

The tidal force can be derived as follows: d is the distance between the two centers of the objects, R is the radius of the planet.

$$F_T = \frac{Gmu}{(d-r)^2} - \frac{Gmu}{d^2} \tag{4.12}$$

using the approximation $r << R$ and $r << d$ we get

$$F_T = \frac{2Gmur}{d^3} \tag{4.13}$$

Then the Roche limit is defined as the point when the gravitational force and the tidal force balance each other: $F_G = F_T$:

$$\frac{Gmu}{d^2} = \frac{2Gmur}{d^3} \tag{4.14}$$

because

$$m = \frac{4\pi \rho_m R^3}{3} \qquad u = \frac{4\pi \rho_u r^3}{3} \tag{4.15}$$

one obtains

Table 4.2: Roche limit for some objects in the Solar System. d_1 denotes the distance for a rigid object, d_2 for a fluid object. r and R are given in units of the main body.

Primary	Object	d_1 (km)	r	d_2 (km)	R
Earth	Moon	9,496	1.49	18,261	2.86
Earth	Comet	17,880	2.80	34,390	5.39
Sun	Earth	554,400	0.80	1,066,300	1.53
Sun	Jupiter	890,700	1.28	1,713,000	2.46
Sun	Moon	655,300	0.94	1,2960,300	1.81
Sun	Comet	1,234,000	1.78	2,374,000	3.42

$$d = r \left(2 \frac{\rho_m}{\rho_u} \right)^{1/3} \qquad (4.16)$$

Roche gave the formula:

$$d \approx 2.44R \left(\frac{\rho_m}{\rho_u} \right)^{1/3} \qquad (4.17)$$

As it has been mentioned, comet Shoemaker-Levy passed within its Roche limit near Jupiter in July 1992. The comet fragmented (Fig. 4.19) and when the comet passed near Jupiter again in 1994, these fragments crashed into the planet.

In Table 4.2 we give the Roche limits for some objects. This Table shows, that *e.g.* a rigid asteroid can approach Earth to a distance of about 17,880 km before disintegrating, whereas a comet can approach to within 34,390 km (assuming it is a fluid).

Saturn's rings are the brightest in the Solar System and the only ones being composed of water ice. The rings of Uranus and Neptune being extremely dark most probably are composed of dark organic material. Jupiter's rings may be made of silicate material.

The moons do not only maintain sharp ring edges but also create the gaps in the rings. For example the 2:1 orbital resonance with Saturn's moon Mimas creates the Cassini division in Saturn's ring system. There are the shepherd moons located close to a narrow ring, one just orbiting inside and the other just orbiting outside the narrow ring.

What is the Roche limit for Earth?

4.4.2 Geologically Active Moons

Io

The innermost of the four Galilean satellites, Io (Fig. 4.20), shows extremely active volcanism. This volcanism is created by the strong tidal forces exerted by the massive planet Jupiter. About 300 volcanic vents and 60 active volcanoes have been found to be known on Io. Sulfurous gases and solids are sprayed up to 300 km above the surface. The surface of Io is very young and colorful. These colors result from mixtures of sulfur, sulfur dioxide and sulfurous salts of sodium and potassium. Io has a diameter of 3643 km and its semi

Figure 4.20: Jupiter's satellite Io; a volcanic plume is seen on the limb. Credit: NASA.

major axis is 421,800 km. One period of revolution about Jupiter takes 1.77 days. As it moves through Jupiter's magnetic field atoms in the higher atmosphere on becoming ionized, escape and are distributed in a torus along its orbit.

Europa

In summary, this moon of Jupiter being similar to our Moon in size however has a surface consisting of water ice that may be melted below, so there is an ocean of liquid water below an ice crust (see Fig. 4.21). Europa being the sixth closest moon of Jupiter was discovered in 1610 by Galileo Galilei. The mean orbit radius is 670,900 km, the orbital period 3.55 days. The mass is 0.008 that of Earth, the mean density 3.01 g/cm^3. The radius is 1,569 km. It is tidally locked to Jupiter like other Galilean satellites.

- Io: 1:1 resonance, one period of revolution about Jupiter equals one period of rotation of Io. The same is valid for the Earth-Moon system. The Moon always shows us the same hemisphere.

- Europa: 2:1 resonance, two rotations = one period of revolution.

- Ganymede: 4:1 resonance

Europa spins faster than it orbits. There must be an asymmetry in internal mass distribution. A layer of subsurface liquid must separate the icy crust from the rocky interior (Fig. 4.22). This subsurface ocean will be discussed in the next chapter.

The total layer of water (frozen at the surface) may be 100 km thick. Because Europa has an induced magnetic field, the water layer must be conductive, as in a salty ocean. On its surface there are only few craters (Fig. 4.23). Like Mars, Europa is a very interesting object

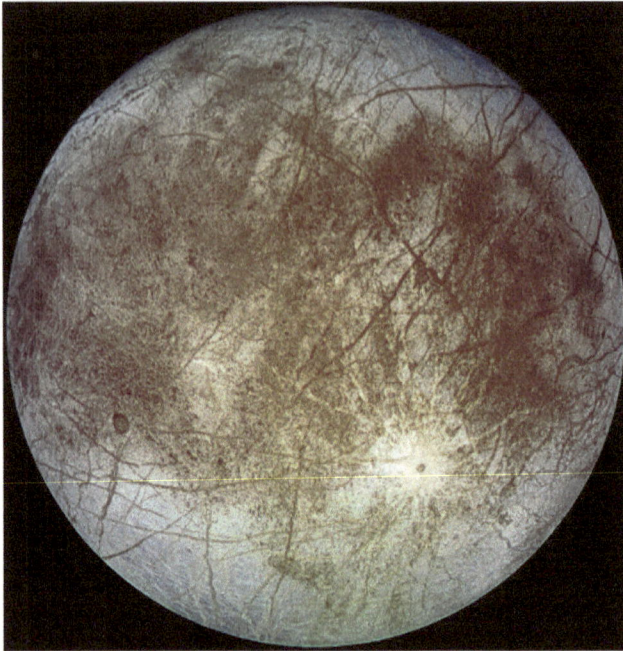

Figure 4.21: Europa. Credit: NASA.

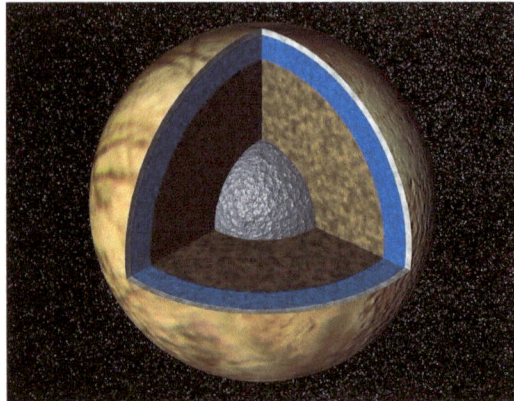

Figure 4.22: Europa: inner structure. Europa has a metallic (iron, nickel) core (shown in gray) drawn to the correct relative size. The core is surrounded by a rock shell (shown in brown). The rock layer of Europa (drawn to correct relative scale) is in turn surrounded by a shell of water in ice or liquid form (shown in blue and white and drawn to the correct relative scale). The surface layer of Europa is shown as white to indicate that it may differ from the underlying layers. Credit: NASA.

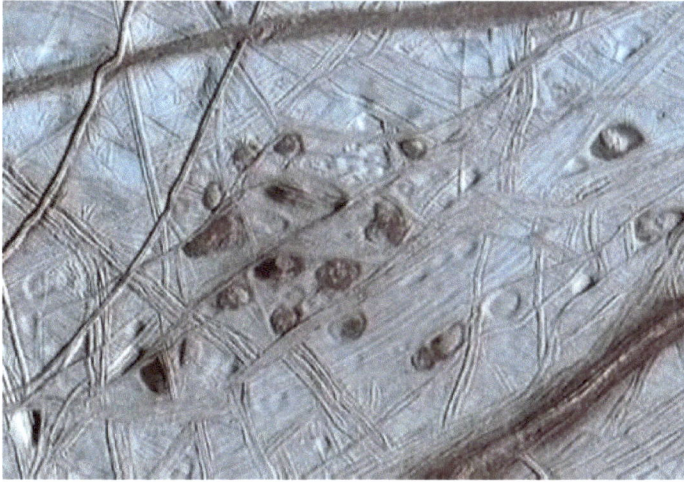

Figure 4.23: Europa, surface structures on the icy crust. Credit: NASA.

for astrobiology. Two important ingredients for life are found there: organic compounds and liquid water.

Why there is liquid water below the ice crust of Europa. The explanation is again tidal force exerted by Jupiter. So the water remains liquid by the tidal heating effect. Since the ocean may be up to 100 km thick, Europa's ocean might contain more water than the Earth's ocean. A strong hint for a subsurface ocean of Europa is the measurements of variations of its magnetic field. This can easily be explained by assuming an internal electrically conducting fluid. Such variations of a magnetic field were also detected on Jupiter's satellite Callisto. This moon could also harbor a subsurface ocean.

Ganymede

Ganymede is the Solar System's largest moon even larger than the planet Mercury. Its semi major axis being 1,070,400 km has a mass 0.025 that of the Earth. It is composed of approximately equal amounts of water ice and silicate rock. A saltwater ocean is believed to exist below a 200 km crust. This satellite being fully differentiated, an iron sulfide - iron core, a silicate mantle and an outer ice mantle, possesses a magnetosphere. Also a faint atmosphere was detected consisting mainly of oxygen. This atmosphere was first noticed during an occultation from ground based observatories and later confirmed the Hubble Space Telescope. HST observations showed an airglow in the far UV that can only be caused by the presence of oxygen.

As it will be discussed later, the presence of oxygen might be used as an indicator for life on a planet. However, in the case of Ganymede, this oxygen is produced by photolysis of the surface water ice. Hydrogen escapes, while oxygen remains in its atmosphere. Ganymede has a low density of $1.9\,\mathrm{g/cm}^3$. Its surface is composed of a dark heavily cratered terrain and a bright terrain with ridges and grooves. Large craters appear shallow. These icy craters deform by viscous flow.

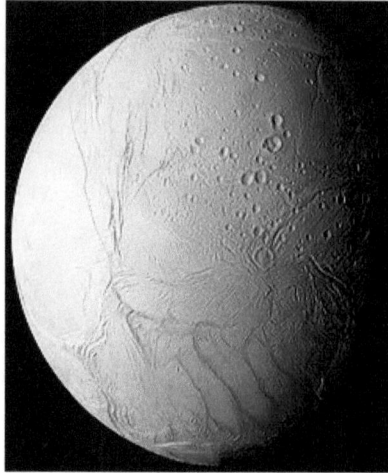

Figure 4.24: Saturn's satellite Enceladus. Credit: NASA/Cassin.

Enceladus

Saturn's moon Enceladus has a diameter of only 500 km. The mean distance to Saturn is about 240 000 km. It shows also evidence of tectonic processes. The surface has an age of only 100 million years. The moon being located in the densest part of the E Ring supplies this ring tiny ice particles those being exhausted from its surface. This water based volcanism on Enceladus is also called cryovolcanism. One still unresolved problem is why Enceladus is so active (Fig. 4.24) as compared to the neighboring satellite Mimas (about the same size). Both of them are certainly subjected to tidal heating caused by Saturn.

A geyser model of cryovolcans at Enceladus is shown in Fig. 4.25.

Titan

Titan is the largest satellite of Saturn and the second largest in the Solar System. It was discovered on March 25, 1655 by the Dutch astronomer Ch. Huygens. Its orbital period being 15 days and 22 hours is identical to its rotational period so Titan is tidally locked in synchronous rotation with Saturn always showing one face to Saturn. The diameter is 5,150 km, being larger than the diameter of Mercury. The density is only 1.88 g/cm^3, so Titan must be composed half of water ice and half of rocky material. The interior may be still hot and liquid consisting of a magma of water and ammonia. This ammonia allows water to remain liquid to temperatures down -97^{0}C. Measurements showed that the crust is decoupled from the interior, providing evidence for a subsurface liquid layer.

Titan being the only known moon with a dense atmosphere has surface pressure of about 1.5 times that of Earth's. The atmosphere consists of opaque haze that blocks the sunlight and prevents from observing the surface directly. We just mention a funny thing: because of dense atmosphere and the low gravity of Titan, humans can fly through it just by flapping wings attached to their arms.

The Huygens probe was released from the Cassini orbiter (orbiting Saturn, launched from Earth 1997) on December 24, 2004 and descended Titan's atmosphere and landed on January 14, 2005. For the first time, direct images of Titan's surface (Fig.4.26) were obtained. Pale hills were detected probably consisting of water ice. The landing site appeared a relatively

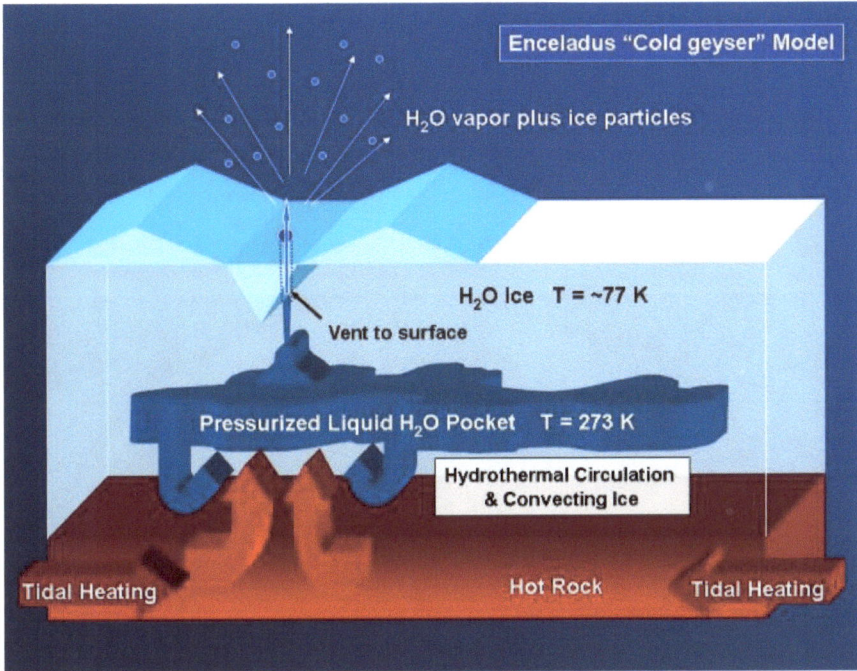

Figure 4.25: Geyser model of the cryovolcans at Enceladus. Credit: NASA/JPL/Space
Science Institute.

flat surface littered with icy rounded rocks.

Titan's surface temperature is about 94 K (-179 ^{0}C). Water ice neither sublimates nor evaporates, since the atmosphere is free of water vapor. Due to its haze, an anti-greenhouse effect is working, reflecting back the sunlight, therefore, making the surface of Titan colder than its upper atmosphere. The clouds are composed of methane, ethane and other compounds. The atmospheric methane creates a greenhouse effect. There is methane rainfall (and other organic compounds) to the surface. There exist lakes of methane which are filled and enlarged depending on seasons.

By synthetic aperture radar (SAR) imaging the surface of Titan was scanned. Signs of volcanism and tectonic activity were found. There are liquid-formed channels.

Triton

Triton is the largest moon in the Neptune system with a diameter of 2700 km (seventh-largest moon in the Solar System). Being in retrograde orbit it is thought to have been captured by Neptune form the Kuiper belt. Its surface temperature being only 38 K consists mostly of frozen nitrogen and also ice crust. The core consists of rock and metal. There are only few craters visible (Fig. 4.27), indicating its surface as geologically young. Voyager 2 found four active geyserlike cryovolcanoes. It seems that nitrogen propels this volcanism. The density of Triton is 2 g/cm^3 and it is composed approximately 15-35% of water ice. Triton has a tenuous nitrogen atmosphere, with trace amounts of carbon monoxide and small amounts of methane near the surface.

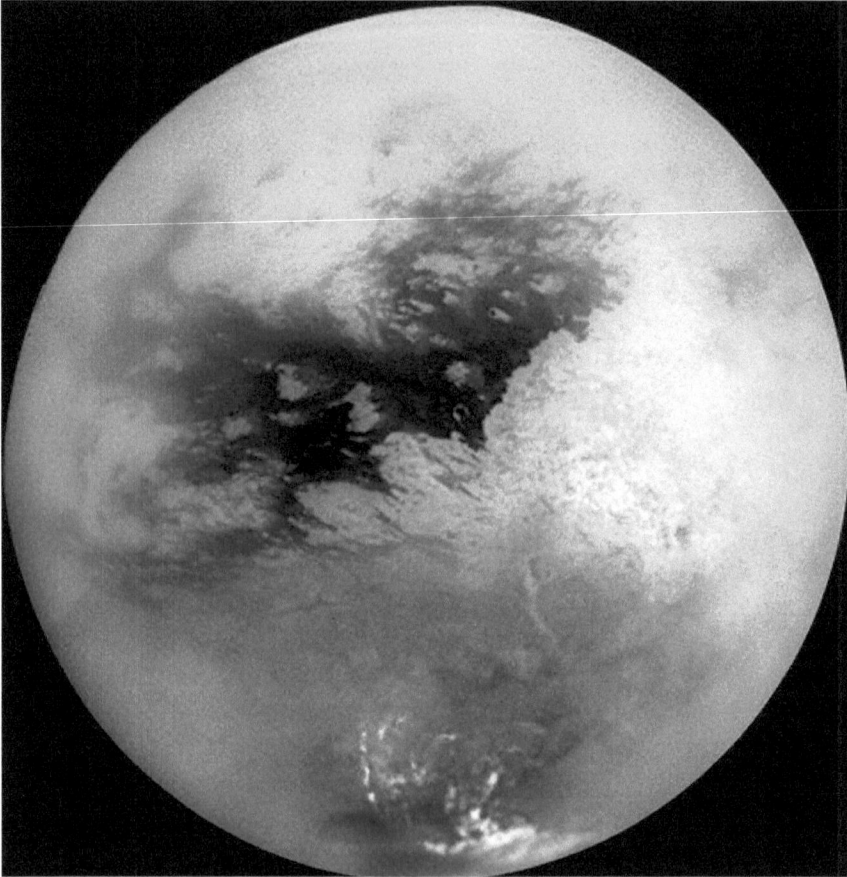

Figure 4.26: The images of Saturn's largest satellite, Titan, reveal a relatively young surface. Credit: NASA.

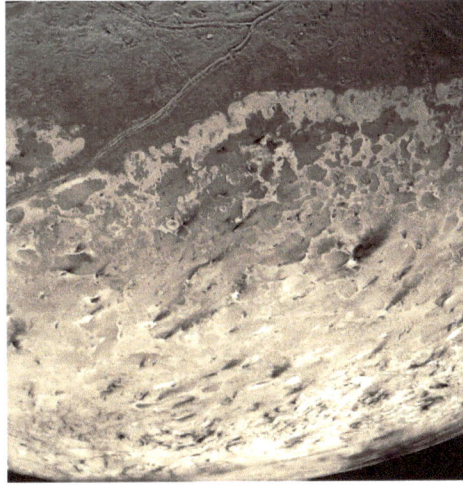

Figure 4.27: Triton's south polar terrain photographed by the Voyager 2 spacecraft. About 50 dark plumes mark what may be ice volcanoes. Credit: NASA.

Activities

Summarize:

- What are internal energy sources of a planet?

- Could there exist a planet which enough internal energy sources to maintain surface temperatures to have liquid water on its surface?

- Discuss similarities between Europa and the Moon.

- Compare the tidal forces on exerted by Earth on Moon, Jupiter on Europa, Jupiter on Ganymede, and Saturn on Titan.

Send Orders of Reprints at bspsaif@emirates.net.ae

CHAPTER 5

Life in the Solar System

Abstract: After having discussed the main bodies in the Solar System, we will review the search for life on planets and their satellites. The most promising candidate to find life is Mars. We review the search for water on Mars, why Mars had undergone drastic climate variations. It is also speculate whether life could exist in the atmosphere of Venus. Besides the planets, life could be also found on several satellites in the Solar System. For astrobiology the Jovian Satellites Europa and Callisto are interesting, probably also the two other Galilean satellites, Ganymede and Io. In the satellite system of Saturn we mention Titan, having a dense atmosphere mainly composed of nitrogen, and Enceladus, where water geysers were observed. The Cassini mission is described in detail. Finally, organic matter found on comets is discussed as well as organic compounds found in planetary disks around stars showing that organic material can be found almost everywhere in the Universe. The main message of this chapter is that in the Solar System the Earth is the only planet where life exists, however, it is still not answered whether life exists on Mars or Europa or even other bodies.

Keywords: Solar System: life; Mars: Life; Mars: Methane; Mars: Water; Venus: Life; Europa: Life; Galilean Satellites: Life; Titan: Life; Cassini Mission; Comets: organic compounds; protoplanetary disks; T Tauri stars

5.1   Venus

5.1.1   Life on Venus Surface

At first glance Venus may seem to present a hostile environment. But this picture may be completely different for early Venus. Four billion years ago, the Sun was 30 % less luminous than today. Oceans could have existed on Venus. It is even possible that the atmosphere of Venus adapted increased solar luminosity providing the oceans the stability quite a long time, maybe even for 2 billion years. Given these facts, there could have been enough time for life to develop on Venus.

Sagan (1967) speculated about life on the surface of Venus (Figs. 5.1, 5.2). At that time it was not known that the surface temperature is almost the same everywhere and that there is no big difference between polar and equatorial regions. He assumed that life might exist in polar regions or on high mountains ([90]).

The evolution of early Venus was reviewed by [20]. Radiogenic noble gas isotopes from the decay of long- and short-lived radioactive parents allow reconstruction of the early and later degassing history of a planet as well as potential early and later atmospheric losses.

Arnold Hanslmeier

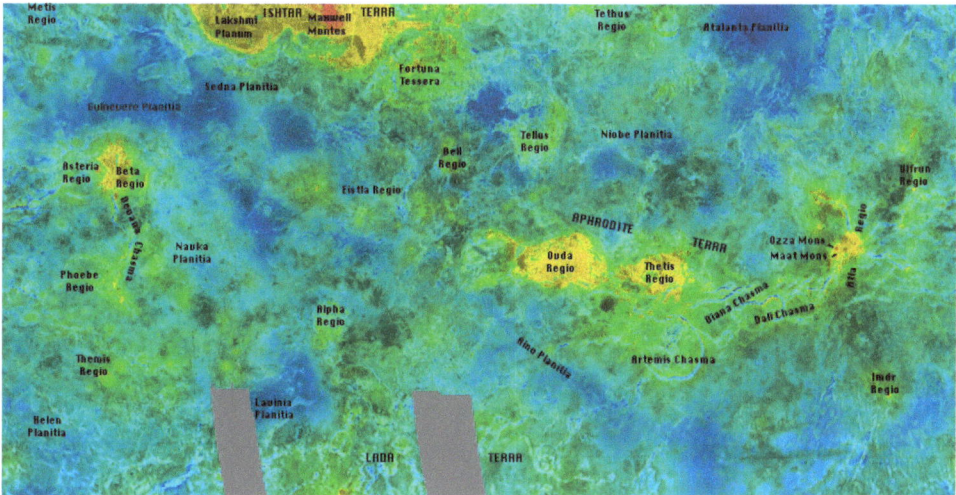

Figure 5.1: Topologic map of Venus, obtained with Magellan Radar. The most notable features are: Ishtar Terra in Northern Hemisphere (about size of Australia Maxwell Montes, highest point on Venus Aphrodite Terra along the Equator (about size of South America) Lack of small craters and big craters in groups.

Argon-40 (from the decay of long-lived ^{40}K likely has been continuously accumulated over >4 billion years and therefore constrains the volcanic/tectonic history of Venus throughout its history.

The radiogenic and fissiogenic fractions of ^{129}Xe and ^{136}Xe [1], produced from now extinct ^{129}I [2] and ^{244}Pu [3], respectively in the atmospheres of Earth and Mars reveal that the two planets underwent a severe early degassing, roughly during the first 100 Myr of Solar System history. For the Earth, this is sometimes attributed to the moon-formation impact only, but data from Mars suggest that a vigorous early degassing may be common to all the terrestrial planets.

At present the surface of Venus seems to be hostile for life.

5.1.2   Life in the Atmosphere of Venus

It has been speculated that life could exist in the high Venusian clouds supporting the idea through two possible hints:

- The carbon monoxide is normally produced by lightening, but the contents of this gas are very low. There exist bacteria and archaea that grow anaerobically utilizing CO as their sole carbon source and water as an electron acceptor to produce carbon dioxide and molecular hydrogen as waste products.

- There is hydrogen sulfide and sulphur dioxide in the Venusian atmosphere, these two compounds normally react and are not found together unless they are being continually produced. Producers of these two compounds could be anaerobic bacteria that decompose organic matter.

---

[1] Half life: 2.11×10^{21} yr
[2] Half life: 1.57×10^{7} yr
[3] Half life: $80{,}0 \times 10^{6}$ yr

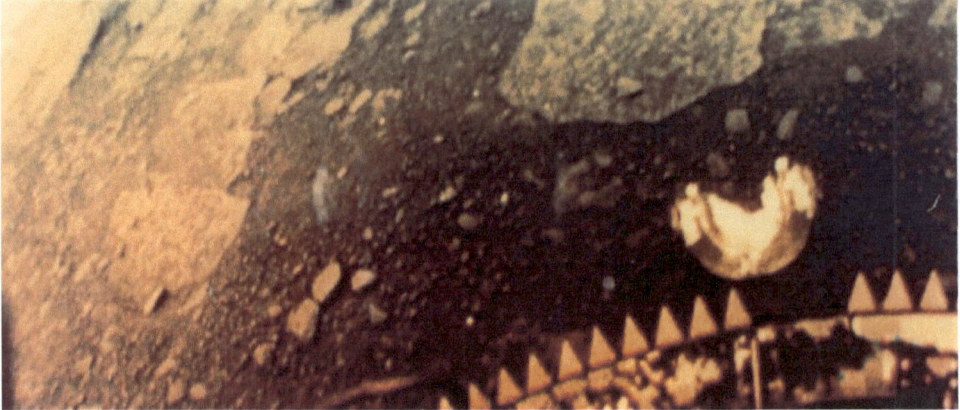

Figure 5.2: Venus: surface around landing site of Venera 13.Credit: IKI.

Thus bacteria could exist like autotrophic thiobacilli, methylotrophs, methanogens, and sulphate-reducing bacteria. They survived the changing climate conditions on Venus and could live at an altitude of about 50 km where the temperatures range from 50 to 70^0 C. At this altitude Venusian clouds are acidic but there are also water droplets.

First speculations on life on Venus were made by [25]. In the vicinity of clouds there is plenty of water, carbon dioxide and sunlight - all that is needed for photosynthesis. At the top of the clouds there maybe ice crystals and at the bottom water droplets. Small amounts of minerals could have been stirred up to the clouds from the surface. This is reviewed in the paper by [75]. One problem how life can survive in the Venusian atmosphere is the strong solar UV radiation. Having originated in a hot proto-ocean or been brought in by meteorites from Earth (or Mars), early life on Venus could have adapted to a dry, acidic atmospheric niche as the warming planet lost its oceans. [95] make the argument that such an organism in the Venusian clouds may utilize sulfur allotropes present in the venusian atmosphere, particularly S_8, as a UV sunscreen, as an energy-converting pigment, or as a means for converting UV light to lower frequencies that can be used for photosynthesis. Thus, life could exist today in the clouds of Venus.

On October 2011, the Venus Express mission found an ozone layer in the atmosphere of Venus. The ozone was detectable as it absorbed some of the ultraviolet rays from the starlight when a star was occulted by Venus. Its ozone layer exists at an altitude of 100 km, being about four times higher in the atmosphere than Earth's and is a hundred to a thousand times less dense. Theoretical work by astrobiologists suggests that a planet's ozone concentration must be 20% of Earth's value before life should be considered as a cause. These new results support that conclusion because Venus clearly remains below this threshold.

Compare the amount of UV radiation received on Top of the Venusian atmosphere with the amount received on top of the Earth's atmosphere.

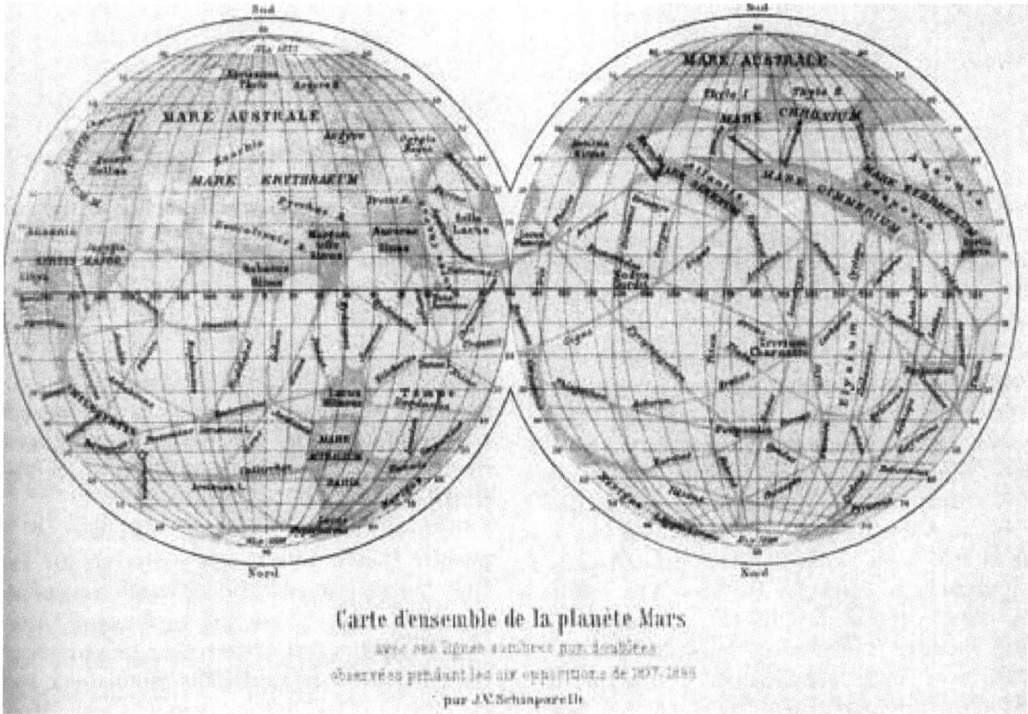

Carte d'ensemble de la planète Mars

avec ses lignes nombres par desilièes

observées pendant les six oppositions de 1877-1888

par J.V. Schiaparelli

Figure 5.3: In 1877, Giovanni Virginio Schiaparelli (1835-1910), began mapping and naming of areas on Mars. He named the Martian "seas" and "continents" (dark and light areas) with names from historic and mythological sources. He saw channels on Mars and called them "canali." As a coincidence, in 1869 the Suez canal was completed, that may have inspired the Mars observers at that time.

5.2   Mars

5.2.1   Early Speculations

The list of speculations about the possibility of life on Mars is a long one. The polar ice caps of Mars were first observed in the mid-17th century. W. Herschel[4] observed their growth and decrease during different seasons on Mars. In 1854 it was assumed that Mars had seas, land and life forms. In 1895 P. Lowell published his book *Mars*, followed by *Mars and its Canals* in 1906. These canals were first seen A. Secchi in 1858 and then by Schiaparelli (see Fig. 5.3) during Mars opposition in 1877. Some observers drew maps of dozens of the canals. E. Maunder conducted visual experiments using volunteers and demonstrated that these canali could be an optical illusion. In his later book Lowell (Fig. 5.4) suggested that the canals were the work of a long-gone civilization. This inspired H.G. Wells[5] to write *The War of the Worlds*. He wrote about an invasion by aliens from Mars who were fleeing the dry planet. In 1909, new telescopic observations under very favorable conditions were made and it became evident, that the canals were an optical illusion.

The first detailed picture of the Martian surface was made by Mariner 4 probe (1965).

---

[4]1738-1822

[5]1866-1947

PLATE VII

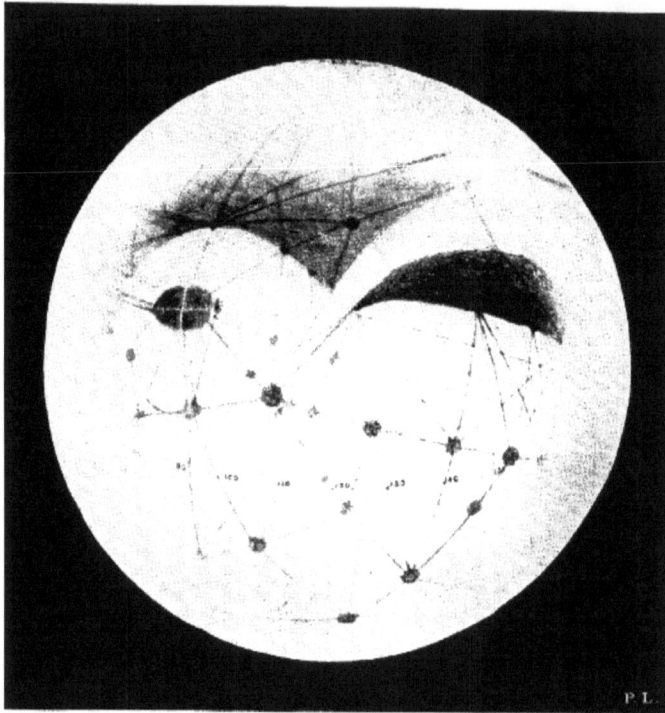

MARS
LONGITUDE 120° ON THE MERIDIAN

Figure 5.4: In 1894, Percival Lowell, a wealthy astronomer from Boston, made his first observations of Mars from a private observatory that he built in Flagstaff, Arizona (Lowell Observatory). He decided that the canals were real and ultimately mapped hundreds of them.

On these images no signs of rivers, oceans or lakes were seen. The surface was covered by craters which indicate the absence of plate tectonics. It was also found that Mars does not have a global magnetic field, so its surface is exposed to cosmic rays.

Then there were several missions to Mars including landers that searched for life on its surface. The Viking probes landed in the mid-1970s, the Phoenix lander in 2008.

5.2.2   Geology of Mars

Studies of impact crater densities on the Martian surface (see Fig. 5.5) showed different periods in the planet's geologic history.

- Pre-Noachian: from the accretion and differentiation of the planet (4.5 billion years ago) to the formation of the Hellas basin (4.1 to 3.8 Gya).

- Noachian Period: formation of the oldest extant surfaces of Mars (between 4.1 and 3.7 Gy ago). These surfaces were scarred by many large impact craters. The Tharsis bulge was formed and large oceans or lakes may have been present. River valley networks were found.

- Hesperian Period: 3.7 to 3.0 Gya. The formation of large extensive lava plains and the formation of the Olympus Mons began as well as ephemeral lakes or seas.

- Amazonian Period: 3.0 Gya to present. In these areas few meteorite impact craters were found. Lava flows, glacial activity and minor releases of liquid water occurred.

Note that the date between the Hesperian/Amazonian Boundary being uncertain could range from 3.0 to 1.5 Gya.

5.2.3   The Surface of Mars

Although all experiments that were conducted so far could not bring evidence for life on Mars, there is still hope to find mircobiological life. However, life on Mars may be hidden like Earth's extremophiles ([53]). It was suggested that the polar regions of Mars are of particular interest for astrobiology exploration. The frozen ground could be a reservoir of organic material protected against atmospheric oxidants. Material between Earth and Mars was exchanged several times. A few dozen of meteorites on Earth were found that came from the Martian surface. Thus, Martian life could be another branch of Earth life. Over the past 10 Myr of Mars three epochs can be found:

- the last 0.5 Myr

- 0.5 to 5 Myr ago

- 5 Myr to 10 Myr ago.

In the first epoch, the formation of melting water appeared possible. In the second epoch, the obliquity of Mars varied over the range 15 to 35^0. The average value is similar to the value today: 25^0. Beyond 5 Myr the obliquity was higher than 35^0. The maximum summer sun was 2.5 times the present value. In Table 5.1, the North Polar Insolation over time is given according to [53].

Life can exist under cold temperature on Mars near polar regions. Imagine ice covered by dust layers[6]. Under such conditions liquid brine solutions may form at temperatures

---

[6]Giant sandstorms on Mars occur frequently

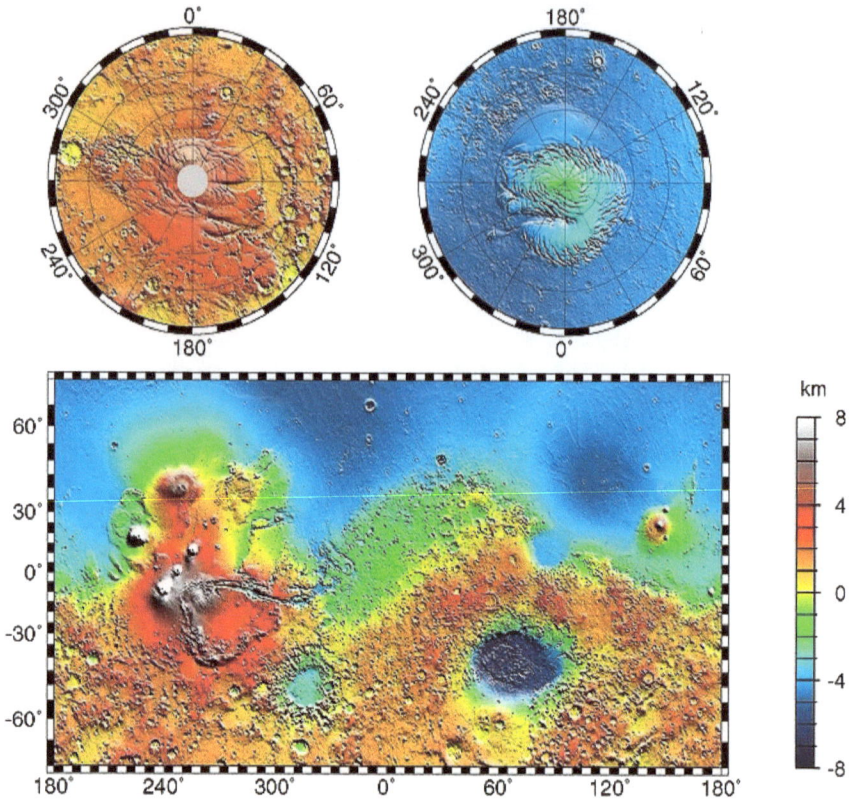

Figure 5.5: Maps of Mars' global topography. The projections are Mercator to 70 latitude and stereographic at the poles with the south pole at left and north pole at right. Note the elevation difference between the northern and southern hemispheres. The Tharsis volcano-tectonic province is centered near the equator in the longitude range 220^0 E to 300^0 E and contains the vast east-west trending Valles Marineris canyon system and several major volcanic shields including Olympus Mons (18^0 N, 225^0 E), Alba Patera (42^0 N, 252^0 E), Ascraeus Mons (12^0 N, 248^0 E), Pavonis Mons (0^0, 247^0 E), and Arsia Mons (9^0 S, 239^0 E). Regions and structures discussed in the text include Solis Planum (25^0 S, 270^0 E), Lunae Planum (10^0 N, 290^0 E), and Claritas Fossae (30^0 S, 255^0 E). Major impact basins include Hellas (45^0 S, 70^0 E), Argyre (50^0 S, 320^0 E), Isidis (12^0 N, 88^0 E), and Utopia (45^0 N, 110^0 E). Credit: NASA/JPL-Caltech.

Table 5.1: Mars: North Polar Summer Insolation (NPSI) over time.

Time period	Max. NPSI
1 Martian year	200 W/m^2
<0.5 Myr	300
0.5 to 5 Myr	400
5 to 10 Myr	500

about -20^0C. Microorganisms could have formed there. [84], considered terrestrial cryogenic ecosystems such as permafrost, overcooled water brines and active volcanoes in permafrost as possible examples for life on Mars.

Besides of the obliquity of its rotation axis large scale volcanism in the Tharsis region could also explain why Mars was wetter in its early history. Volcanoes exhaust CO_2 which is a greenhouse gas and therefore the Martian surface was warmer. Also water vapor is exhausted during eruptions and it is estimated that enough water water was exhausted to cover Mars globally with an ocean 120 m of thickness.

A recent simulation done by [37] arrived at the conclusion that significant climate change in the recent past on Mars can also be explained a cloud greenhouse effect. Summarizing, climate changes on Mars can be explained by

- Variations of rotation axis obliquity,

- Cloud greenhouse effect,

- Volcanism

Compare possible life on Mars near the poles with life on Earth in artic/antarctic regions.

Discuss how the above mentioned explanations of climate change on Mars can interfere.

5.2.4 The Martian Face

One of the features in the Cydonia region, the "face on Mars" (about 1.5 kilometers across), appears like a face (Fig. 5.6). On July 25, 1976 the probe Viking 1[7] took an image that appeared like a humanoid face. This attracted great interest both in the scientific community as well as in the science fiction community. After more than 20 years of painstaking struggle spacecrafts like NASA's Mars Global Surveyor (1997-2006) and Mars Reconnaissance Orbiter (2006-) and the European Space Agency's Mars Express probe (2003–) observed this region. In contrast to the relatively low resolution of the Viking images of Cydonia, these new platforms afford much improved resolution. For instance, the Mars Express images are at a resolution of 14 m/pixel or better. By combining data from the High Resolution Stereo Camera (HRSC) on the Mars Express probe and the Mars Orbiter Camera (MOC) on board NASA's Mars Global Surveyor it has been possible to create a 3D representation of the "Face on Mars". These high resolution images showed that the Face on Mars is an optical illusion (Fig. 5.6). The Face on Mars was also discussed by [80].

5.2.5 Meteorites from Mars

There are several meteorites found on Earth that come from Mars. The ALH84001 meteorite (Fig. 5.7) was found on December 1984 in Antarctica. It weighs 1.9 kg and was ejected from Mars about 17 million years ago. It spent 11,000 years in the Antarctic ice sheets. It was found that 25% of the magnetite found in this sample occurs as small crystals that on Earth,

---

[7]Launched Aug 1975; Viking 1 was the first of two spacecraft sent to Mars as part of NASA's Viking program. It was the first spacecraft to successfully land on Mars and to perform its mission, and until May 19, 2010 held the record for the longest Mars surface mission of 6 years and 116 days (from landing until surface mission termination, Earth time).

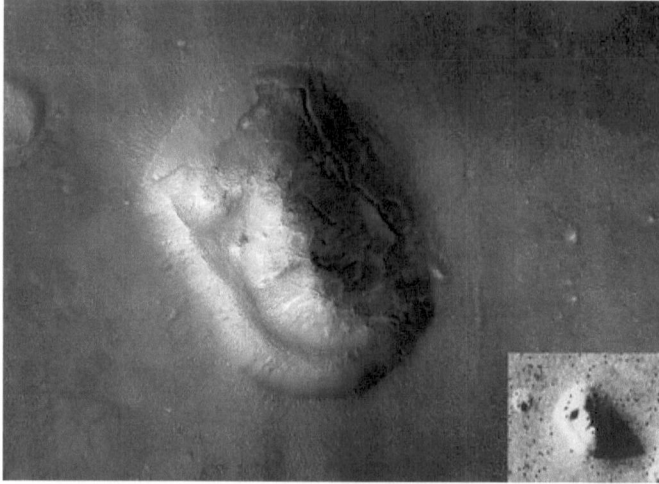

Figure 5.6: Mars Reconnaissance Orbiter image by its HiRISE camera of the "Face on Mars" Viking Orbiter image inset in bottom right corner. Credit: NASA / JPL / Malin Space Science Systems

is associated with biological activity. Also polycyclic aromatic hydrocarbons were identified. The conclusion was, that this meteorite provides strong evidence that life may have existed on ancient Mars. Elongated prismatic magnetite crystals in ALH84001 carbonate globules were considered as evidence of potential Martian magnetofossils [103]. Geochemistry of the ALH84001 was revised by [3] and later, the biological interpretation for ALH 84001 became strongly disfavored.

The Nakhla meteorite fell on Earth in 1911. Structures that resembled fossilized bacteria on Earth were found but it is still controversial whether these can be really interpreted as signs of biologic activity on Mars.

The 4 kg Shergotty meteorite fell on Earth in 1865. It was formed on Mars 165 million years ago. The results concerning life on Mars are still controversial.

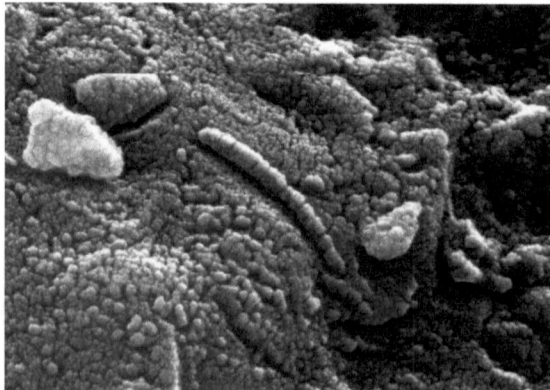

Figure 5.7: Structures on ALH84001 meteorite. Credit: NASA

Also meteorites on the surface of Mars have been detected by Mars exploration rovers ([32]). These seem to support the hypothesis that Mars once had a denser atmosphere and considerable amount of water and/or water ice near the surface or even at the surface. This was concluded since the meteorites found underwent significant chemical weathering due to aqueous alteration. The cavernous features on the surfaces of the meteorites suggest differential acidic corrosion removing less resistant material and softer inclusion.

5.2.6 Water on Mars

Hematite is found on the surface of Mars. This mineral forms in the presence of water. Hematite is a mineral, black to steel or silver-gray, brown to reddish brown, or red. Chemically, it is an iron (III) oxide:

$$Fe_2O_3 \tag{5.1}$$

It can be precipitated out of water in layers (such at the bottom of a lake), spring or standing water. Without water, hematite results in volcanic activity.

The Mars exploration rovers detected hematite spherules embedded in the rock. These spherules were named blueberries.

Under present conditions, liquid water cannot exist on Mars: the temperature and the pressure in its atmosphere are too low. In 2000 flood-like gullies were discovered. This provided evidence for water under the surface of Mars. Such sub surface water reservoirs might form a present-day habitat for life. Water ice was discovered at the south pole of Mars in 2004 and also evidence for water ice near the north pole of Mars was found one year later. Images from the Mars Global Surveyor Mission suggested that water occasionally flows on the surface of Mars. Craters and sediment deposits changed over only a few years. Analysis of Martian dust storms showed that if there is water on Mars it would be extremely salty, perhaps too salty for life. But we know from extremopohiles on Earth such as Haloarchaea being able to live in hypersaline solutions.

Discuss the different evidence for water on Mars? Why can water not exist in liquid form at its surface under present conditions?

5.2.7 Methane on Mars

Methane (CH_4) has been detected in the Martian atmosphere at a global mixing ratio of 10- 15 ppbv with spatial and temporal variations ranging between 0 and 45 ppbv (see Fig. 5.8). Due to photocatalytic processes in the upper atmosphere, methane is unstable, with the estimated lifetime being about 300 years and therefore it must be replenished in the atmosphere of Mars, producing about 270 t/yr. So there must be a source for this gas. Asteroid impacts account for less than 1% of the total methane production. There is low geologic activity on Mars. Methane production by volcanism can be excluded. Thus there are speculations that microorganisms such as methanogens produce methane on Mars. Methanogens are microorganisms in the domain Archaea those can metabolize H_2 as an energy source, CO_2 as a carbon source, and consequently produce methane. Methanogens have been shown to produce methane at reduced pressures (400 and 50 mbar), in the presence of perchlorate salts, using carbonate as a sole carbon source, and to survive desiccation at both 1 bar and 6 mbar for extended periods of time ([54]). The problem for such organisms

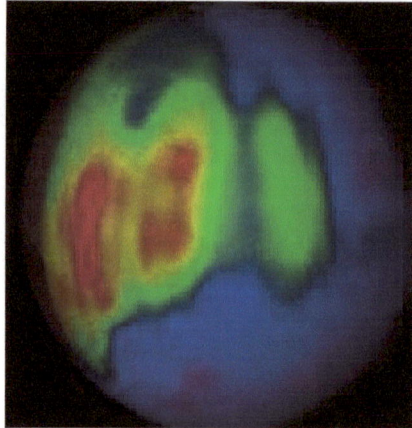

Figure 5.8: Methane in the atmosphere of Mars. Credit: NASA

is the dry surface of Mars. The survival of methanogens during desiccation was tested by laboratory experiments by [49]. Methanosarcina barkeri survived desiccation for 10 days, while Methanobacterium formicicum and Methanothermobacter wolfeii were able to survive for 25 days.

Methane plumes were found which started to show up in the northern hemisphere spring of Mars, gradually building up and peaking in late summer. A plume contained about 19,000 metric tons of methane. The UV/CH_4 model for Mars is now supported by four studies that demonstrate the evolution of CH_4 from UV-irradiated organics under simulated Martian conditions. The UV/CH_4 model predicts a global average up to 11 ppbv[93].

In May 2007, the Spirit rover found an area extremely rich in silica (90%). This is reminiscent of the effect of hot spring water or steam coming into contact with volcanic rocks. Spider like radial channel carved on 1 m thick ice are formed near the Southern pole, which results from the seasonal frosting and defrosting processes. When winter gets over, they sublime and CO_2 and water geysers may form. In July 2012, the Mars rover Curiosity successfully landed. This car sized rover will perform several experiments near the crater Gale.

Compare the amount of UV flux at the top of the atmosphere of Mars with that received by Earth at the top of its atmosphere.

5.3   The Galilean Moons

The four largest satellites of Jupiter, Io, Europa, Ganymede and Callisto are called the Galilean Moons since they have been first mentioned by Galileo Galilei.

5.3.1   Europa

As it has been discussed in the previous chapter, this satellite has a surface ice crust and beneath this crust a liquid salty ocean is assumed.

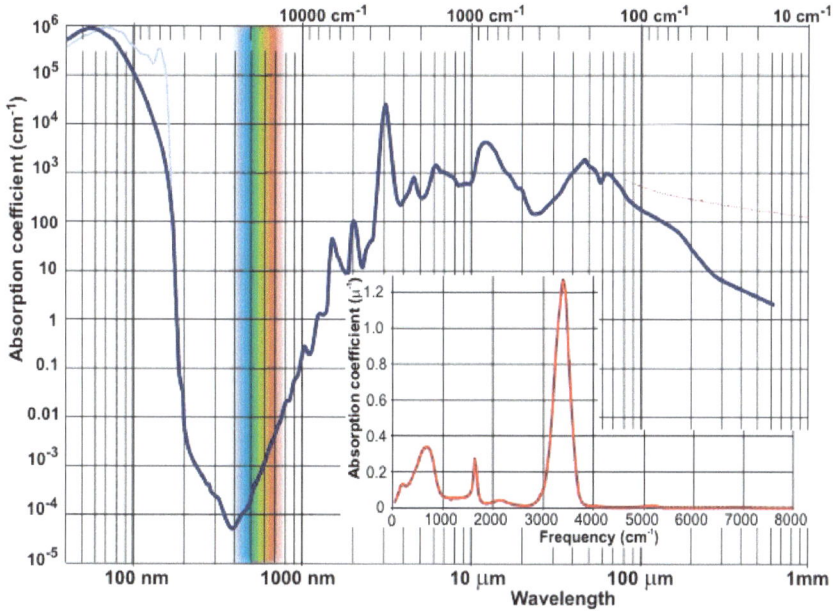

Figure 5.9: Water spectral signatures. In the visible the absorption coefficient is low, so that the Earth's atmosphere is strongly transparent. Note the strong absorption in the IR. Credit: Martin Chaplin, 2011.

Europa has a tenuous O_2 and O atmosphere. With the HST two FUV oxygen lines were detected at 130.4 and 135.6 nm, respectively [91]. The non-uniformity of the O_2 distribution was discussed in [17].

Water ice on the surface of Europa has been discovered by means of ground-based astronomy long before the Voyager spacecrafts. The albedo is 0.64 which is explained by a surface of high reflectivity. Kuiper measured already in 1957 infrared spectra of Europa and found a depression at $\lambda > 1.5\,\mu m$. This feature seems to indicate the presence of snow ice on the surface of Europa ([56]). Near-infrared spectra of Europa and Ganymede show an H_2O ice band feature near $1.5\,\mu m$ (see Fig. 5.9).

The dayside temperature was estimated at 120-135 K. The first direct images of Europa were obtained in 1979 during Voyager 1 and 2 flybys. They revealed a smooth bright surface with low density of impact craters and a network of dark linear structures extending over thousands of kilometers. In 1995, the Galileo spacecraft entered into the orbit of Europa and high resolution images of Europa showing details up to several meters were made. Since Europa is in strong tidal interaction with Jupiter and other satellites, tectonic activity is high and viscous material extrudes to the surface like the middle oceanic ridges as a result of repeated magma flow on Earth.

The radiation level on Europa's surface is about 540 rems per day. This amount may cause illness or death in human beings. However, beneath the ice crust there is protection against this radiation. Dark streaks found on its surface may be explained by eruptions of warm ice, the crust spreads open.

The surface is younger than 100 million years. The most convincing evidence of a subsurface global ocean comes from Galileo magnetic field measurements. Because of the rotation of Europa, a magnetic field is induced (since currents flow in the salty subsurface ocean).

That induced magnetic field is variable in direction and strength depending on Europa's position within Jupiter's magnetic field. The data require a complete spherical shell of salty water and the estimates of the ice-shell thickness range from several 100 m to more than 30 km.

Therefore, this satellite is an interesting target for future space missions. If there does exist life in the subsurface ocean of Europa a special strategy to search for this life has to be developed. This includes a careful analysis of endogenous materials that arise from the interior.

The subsurface ocean environment of Europa may be similar to that of deep ocean hydrothermal vents on Earth. But there is also an important difference. On Earth, the existence of life at great depths in the oceans is not a proof that life has originated there. Microorganisms from Earth might have been transported to Mars, so an insemination has occurred there. Assuming that meteorites that hit the surface of Earth have delivered prebiotic organics to early Earth or possible inseminated Mars, this mechanism will not work for Europa, giving the possibility of life on Europa:

- to have been originated in the depth of subsurface ocean,

- was independent from the energy from the Sun,

- originated independently of the life on Earth.

[82], suggested strategies to search for life on Europa through simple physico-chemical traces of metabolism to complex biomolecules or biostructures. Raman spectroscopy or biosensor technologies are the future for *in situ* exploration of the Solar System.

If biologically useful substances are deposited by bombardment to Europa's surface they must be delivered down through the icy crust, making the ice layer oxygenated. The search for life on Europa was also discussed in [39] and [40].

LAPLACE is a mission to Europa and the Jupiter System. It is in the European Space Agencie's Cosmic Vision Programme, with the running title now being Europa Jupiter System Mission (EJSM-Laplace). The cruise to the Jupiter system will last for 6 years and it will cruise for three years in Jupiter's system. Included in the mission is a JGO, Jupiter Ganymede Orbiter. The launch of this mission is due by 2020.

Detection of life in the subsurface ocean of Europa will be a major step forward in our understanding of life. Until the 1970s it was believed that life depends on energy from the Sun. In 1977, the deep-sea exploration submersible Alvin discovered life (worms, clams, crustaceas) around undersea volcanic features known as black smokers.

5.3.2   Callisto

The surface of Jupiter's moon Callisto (Fig. 5.10) is heavily cratered with frost deposit indicating that the surface is very old. A tenuous carbon dioxide atmosphere on Callisto was first reported by [15]. An off-limb scan was performed by the Galileo spacecraft during a close flyby of Callisto. Data from an IR spectrometer were analyzedhighlighting that Callisto is surrounded by an extremely thin atmosphere composed of carbon dioxide and probably molecular oxygen, as well as by a rather intense ionosphere. The surface pressure is estimated to be 7.5×10^{-12} bar and the surface particle density is 4×10^8 cm^{-3}. Because of atmospheric escape, such a thin atmosphere would be lost in only about 4 days, so it must be constantly replenished. One suggested mechanism is slow sublimation of CO_2 ice from the satellite's icy crust.

There exists a salty ocean beneath the surface of Callisto. Evidence for subsurface oceans in Europa and Callisto are induced magnetic fields ([52]). However, there may be

Figure 5.10: Callisto. Prominent features on its surface are multi ring structures. Credit: NASA/JPL.

some important differences between Callisto and Europa making life less probable there. There is a lack of contact with rocky material and the heat flux from the interior of Callisto is lower than for Europa. The difference between Europa and Callisto concerning the liquid ocean might be the heating mechanism:

- Europa: heating by tidal interacting with Jupiter.

- Callisto: heating mainly by radioactive decay.

The lower tidal forces on Callisto are explained by the larger distance to Jupiter: the mean distance being about 1,882,000 km is about 3 times the distance of Jupiter–Europa. Since the tidal forces depend on r^{-3}, r being the distance of the satellite to its parent planet, the tidal forces on Callisto are about 30 times weaker than those on Europa.

Organics and other molecules in the surfaces of Callisto and Ganymede were discussed by [73].

The most prominent feature on Callisto's surface is Valhalla (Fig. 5.11), possibly created by a large asteroid or comet which impacted Callisto. Valhalla consists of a bright inner region, about 600 kilometers in diameter surrounded by concentric rings 3000 to 4000 kilometers (1800-2500 miles) in diameter. The bright central plains were possibly created by the excavation and ejection of "cleaner" ice from beneath the surface, with a fluid-like mass (impact melt) filling the crater bowl after impact. The concentric rings are fractures in the crust resulting from the impact.

[62] gives a review on Callisto's atmosphere.

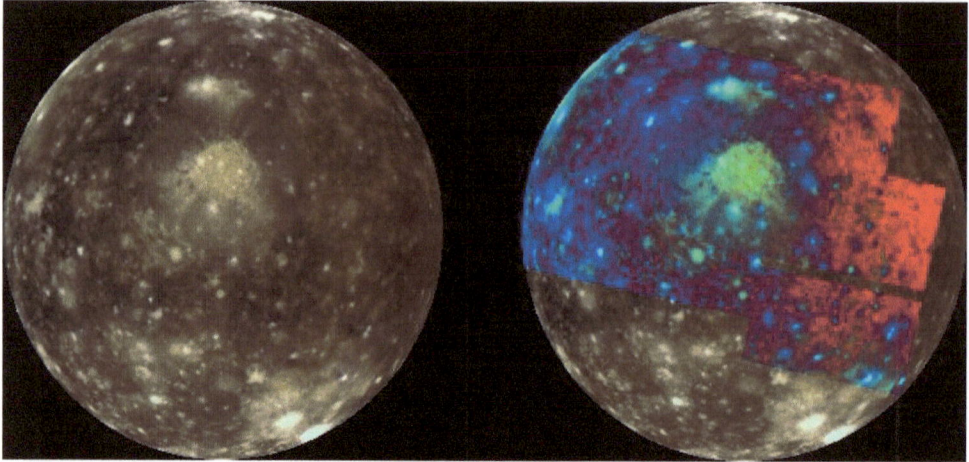

Figure 5.11: Callisto. At the North of the equator the impact bassin Valhalla is seen. The color image on the right was obtained with the 1 micrometer (infrared), in green and violet filters of the Solid State Imaging (SSI) system on NASA's Galileo spacecraft. It reveals a gradual variation. Credit: NASA/JPL/University of Arizona

5.3.3 Io

Io is the most active volcanic object in the Solar System at present because of its strong tidal interaction with Jupiter. The semi major axis is larger than that of our Moon, being only about 421 000 km. Plumes were observed on the bright and dark limbs of the satellite corresponding to half of the surface ranging from 10 to 75 km in width and 280 to 70 km in height[101]. Jupiter possesses a very strong magnetic field (Fig. 5.12) which sweeps up gases and dust from Io's thin atmosphere. Therefore, Io loses about 1 ton per second. Io has an extremely thin atmosphere consisting mainly of sulfur dioxide (SO_2), with minor constituents including sulfur monoxide (SO), sodium chloride (NaCl), and atomic sulfur and oxygen (see also [61]).

The possibility of life between fire and ice on Io was mentioned by [94].

Since there is much sulfur on Io we can ask the question whether life based on sulfur could exist there. Sulfur-dependent life on Earth releases hydrogen sulfide as waste. These microbes could increase hydrogen sulfide levels by nearly 10 times what they would be on a planet without such life. Thus the detection of sulfur on an extrasolar planet can be interpreted as having been produced by life based on sulfur. Microbes can live off the energy available in sulfurous molecules released by volcanoes essentially "breathing" these compounds the way humans breath oxygen. The sulfur bacteria known on Earth use mainly the two substances hydrogen sulfide, H_2S and thiosulfate S_2O3^{2-} for their metabolism.

Purple sulfur bacteria and green sulfur bacteria utilize energy from light in an oxygen-free environment transforming sulfur and its compounds to sulfates.

Life at Io's surface is highly improbable because there is no protection against radiation and particles (these are also triggered by Jupiter's strong magnetic field). The interior of Io seems to be hostile to life because Galileo magnetometer data in 2009 revealed the presence of an induced magnetic field at Io, requiring a magma ocean 50 km below the surface ([50]).

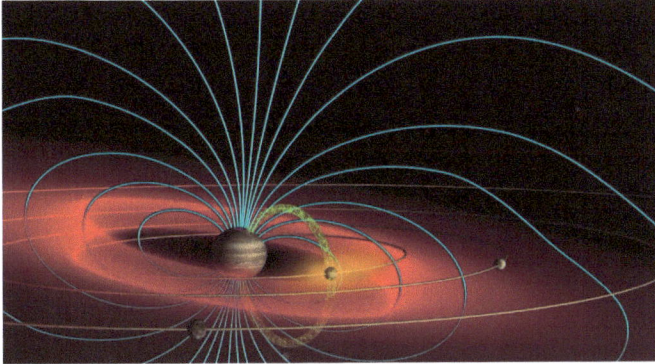

Figure 5.12: The magnetosphere of Jupiter and the plasma torus of its inner satellite Io.
Source: wikimedia.

Why is life based on sulfur metabolism interesting for the search of life on Io? Could life be expected in the interior of Io?

5.3.4 Ganymede

This satellite (Fig. 5.13) is larger than planet Mercury but it contains only 45% of Mercury's mass. Composed of silicate rocks and water ice, Ganymede is the only satellite in the Solar System known to possess a magnetosphere, likely created through convection within the liquid iron core. The weak magnetosphere is buried within Jupiter's much larger magnetic field and connected to it through open field lines. The satellite has a thin oxygen atmosphere that includes O, O_2, and possibly O_3. Europa and Ganymede are imbedded in the Jovian magnetospheric plasma. The incident plasma alters the surface and produces tenuous atmospheres often called surface boundary layer atmospheres ([46]). [51] reported on magnetic induction signals at Ganymede as implications for a subsurface ocean. There must be a molten iron core which also implies a molten subsurface ocean and makes Ganymede an interesting object for the search for life.

5.4 Life in the Saturnian System?

5.4.1 Titan

Titan is the largest satellite of Saturn and is the second largest satellite in the solar system.

Titan has long been a subject of interest. It provides an excellent example of abiotic processing of organic material. Titan is the only satellite known in the Solar System to have a dense atmosphere. Its atmosphere was detected by J. Campos Sola from measuring the limb darkening in 1903. 98.4% of its atmosphere is nitrogen, the remaining 1.6 % being composed mainly of methane, CH_4 and other trace gases. Therefore, Titan and Earth are the only two

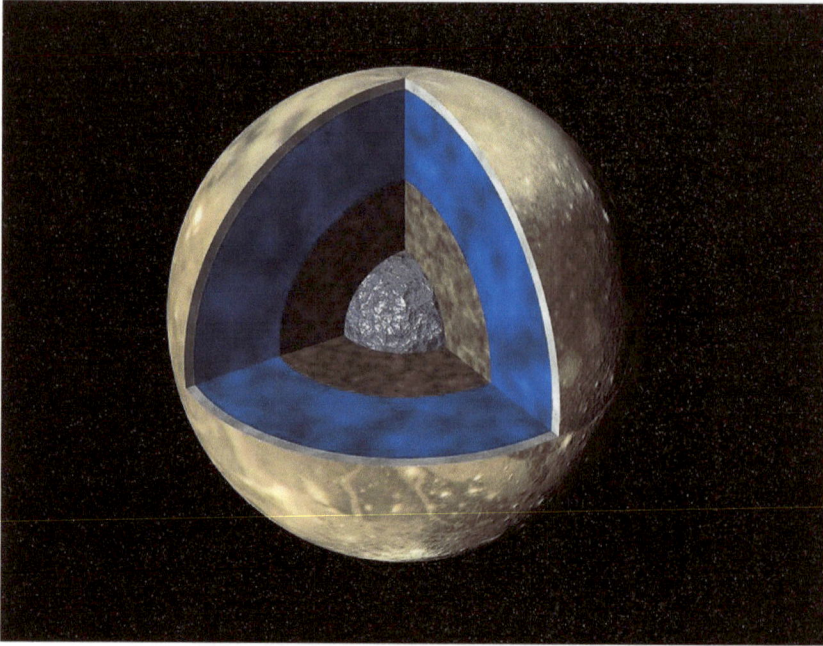

Figure 5.13: Interior of Ganymede.

objects in the Solar System that contain a nitrogen rich atmosphere. Among the trace gases we find ethane, diacetylene, methylacetylene, acetylene, propane, cyanoacetylene, hydrogen cyanide, carbon dioxide, carbon monoxide, cyanogen, argon and helium.

Photochemical reactions produce complex organics involving large amounts of hydrogen. In these processes, EUV and VUV radiation are involved. Tholin (after the Greek word for muddy), is a heteropolymer formed by solar ultraviolet irradiation of simple organic compounds such as methane or ethane. Tholins do not form naturally on modern-day Earth, but are found in great abundance on the surface of icy bodies in the outer Solar System. They usually have reddish-brown appearance.

The orange color of its atmosphere is produced by chemicals such as tholins (see Fig. 5.15) being tar-like substances. Hydrocarbons are formed in Titan's higher atmosphere by photolysis of methane by solar UV radiation thus producing thick organic orange smog. It is assumed that methane would transform in about 50 million years into more complex hydrocarbons. Thus there must be a replenishment of methane. It is believed that this methane reservoir, that cannot be attributed to cometary impacts (because comets contain only small amounts of methane) is released via volcanic eruptions.

By diameter and volume Titan is larger than planet Mercury, but only half as massive. This implies that this planet like satellite is mainly composed of rock and ice. There have been many speculations about the nature of its surface. It was suspected that there might be surface structures similar to that found on Earth.

Four NASA spacecrafts have been sent to explore Saturn. Pioneer 11 was first to fly past Saturn in 1979. Voyager 1 flew past one year later, followed by its twins, Voyager 2, in 1981. The Cassini spacecraft launched on October 15, 1997 was the first to explore the Saturn system of rings and moons from orbit. It is interesting to mention the complicated cruise to Saturn: Cassini performed two gravity-assist flybys of Venus on April 26, 1998 and June

Figure 5.14: Lakes of liquid ethane, methane and nitrogen on the surface of Titan. Credit: Cassini-Huygens mission.

Table 5.2: Haze production on Titan and the early Earth.

	Titan	Earth
l_0	6.2×10^9	1.4×10^{12}
F, aerosol production flux $\mathrm{gcm^{-2}s^{-1}}$	1.2×10^{-14}	8×10^{-11}
Aerosol production $\mathrm{gyr^{-1}}$	8×10^{10}	3×10^{15}

24, 1999. On August 18, 1999 Cassini did a gravity-assist flyby of Earth. An hour and 20 minutes before closest approach, Cassini made the closest approach to the Moon at 377,000 km, and took a series of calibration images. On Jan. 23, 2000, Cassini performed a flyby of Asteroid 2685 Masursky. Cassini entered the orbit about Saturn on Jun. 30, 2004. The Cassini orbiter was built and managed by NASAs Jet Propulsion Laboratory. The Huygens probe was built by the European Space Agency. The Huygens probe separated on December 25, 2004 and successfully descended to Titans surface on January 15, 2005. It landed on solid ground with no liquids in view. Although Cassini successfully relayed 350 of the pictures that it received from Huygens of its descent and landing site, a software error failed to turn on one of the Cassini receivers and caused the loss of the other 350 pictures. Radar images obtained on July 21, 2006 appear to show lakes of liquid hydrocarbon (such as methane and ethane) in Titans northern latitudes. This is the first discovery of currently-existing lakes anywhere besides Earth. The lakes range in size from about a kilometer to one which is one hundred kilometers across.

The Cassini Radar flyby of Saturn's moon Titan on 22 July 2006 provides compelling evidence for the presence of liquid lakes on the surface of Titan (Fig. 5.14). The radar images polewards of 70^0N show over 75 circular to irregular radar dark patches from 3 km to over 170 km across, in a region where liquid methane and ethane are expected to be abundant and stable on the surface ([68]). Thus the Cassini mission proved the existence of liquid lakes on Titan's surface. The Cassini-Huygens probe flew by in February, 2007, taking radar and camera pictures of some large areas near the north pole that may be large areas of liquid methane and/or ethane, including one sea larger than 100,000 $\mathrm{km^2}$ (larger than Lake Superior), and another area similar in size to the Caspian Sea[8]. Pictures taken near Titan's southern pole in October 2007 showed similar, but far smaller, areas that look like lakes.

In Table 5.2 (adapted from [105]) the haze production on Titan and the early Earth is given. l_0 is calculated for Earth by multiplying the present solar EUV flux with 2.5 to account for the stronger EUV emission for the early Sun.

The moment of inertia of an object can be estimated from its quadrupole gravity field. Titan's moment of inertia was found to be 0.32 from Cassini measurements. This is equally consistent with a fully differentiated internal structure comprising a shell of water ice overlying a low-density silicate core; depending on the chemistry of Titans subsurface ocean, the core radius is between 1980 and 2120 km ([33]). The interior structure of Titan is shown in Fig. 5.17.

Compare Titan with Europa. What are similarities and dissimilarities? Where could life exist on these satellites?

---

[8]Surface area 371,000 $\mathrm{km^2}$

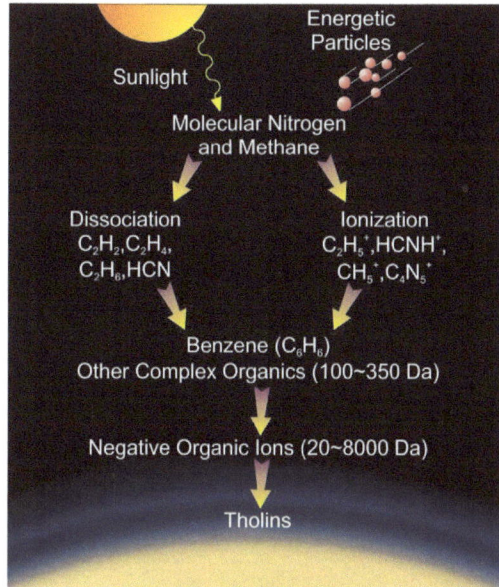

Figure 5.15: Tholin formation in Titan's atmosphere. Credit: Southwest Res. Inst.

5.4.2   Enceladus

Enceladus (Fig. 5.18) was discovered in 1789 by W. Herschel. This Saturnian satellite is strongly influenced by Saturn's E-ring. The Cassini orbiter showed many details on its surface that reminds to the surface features on Jupiter's satellite Europa, therefore assuming that also Enceladus could harbour a subsurface ocean. In 2005, Cassini detected plumes of water rising and the resulting ejected ice particles could be measured. Five different types of terrain are found on its surface, including several regions of cratered terrain, regions of smooth, young terrain, and lanes of ridged terrain that often borders the smooth terrain. Within the smooth plains there are no craters. These regions are relatively young, probably less than 100 million years. This means that Enceladus must have been active very recently with some sort of "water volcanism" or other processes those renewing the surface. The fresh, clean ice that dominates its surface gives Enceladus the highest albedo of any body in the Solar System (Visual geometric albedo of 0.99). Because it reflects so much sunlight, the mean surface temperature is only -201^{0}C. Tectonic features found on Enceladus are rifts that can run for up to two hundred kilometers long, 5–10 kilometers wide, and are up to a kilometer deep. The magnetometer instrument on the Cassini spacecraft may have detected an atmosphere at Enceladus. The magnetometer observed an increase in the power of ion cyclotron waves near Enceladus. These waves are produced through the ionization of particles within a magnetosphere and the frequency of the waves can be used to identify the composition, in this case ionized water vapor (Burton 2005). As Enceladus's gravity is too weak to retain an atmosphere, it must be replenished from some source, possibly from icy volcanoes or geysers. Cassini discovered water vapor plumes ejected from the south pole. These plumes may extend up to 80 km above the surface due to the moon's low surface

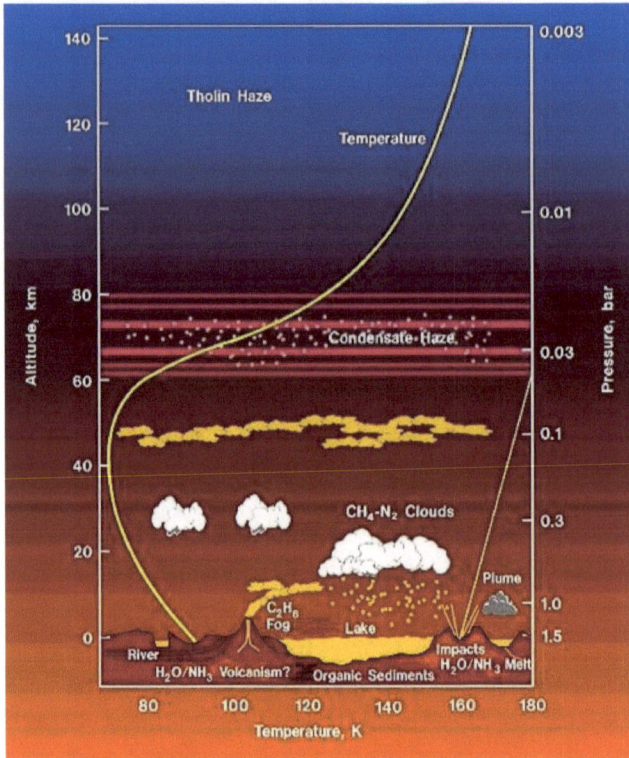

Figure 5.16: Structure of Titan's atmosphere. Credit: John Baez

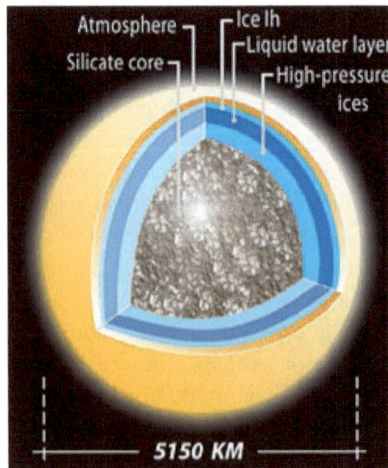

Figure 5.17: Interior of Titan. Credit: NASA

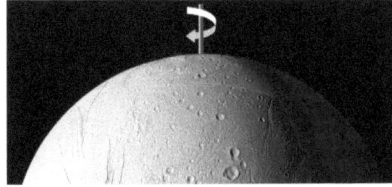

Figure 5.18: Enceladus and cutaway. Credit: Cassini Mission

gravity.

If there is CO, CO_2 and NH_3 present in the spectra obtained from the plume, then there is possible evidence that amino acids could be formed at the rock/liquid interface of Enceladus. The combination of a hydrological cycle, chemical redox gradient and geochemical cycle gives favorable conditions for life ([22]).

Europa, Callisto, and Ganymede in the Jovian system, and Enceladus, Titan, Hyperion, Iapetus, and Hyperion in the Saturnian system all possess characters that could make them conducive to the origin or maintenance of life upon or within them. The possibility of some of these bodies containing extraterrestrial life is reflected in future explorative missions [6].

Which of the Jovian or Saturnian moons would you consider to be the most probable candidate for life to exist on it?

5.5   Comets and Asteroids

Already in the 17th century the possibility that comets may have played an important role in the formation of planets, their atmospheres and evolution of life on Earth was proposed by E. Halley and I. Newton. They believed that comets have delivered water to the atmosphere and the ground. The nucleus of a comet consists of ice-rich material.

5.5.1   Comets and Water on Earth

Comets may have contributed to the volatile elements on Earth. As the other terrestrial planets, Earth accreted from dust particles. Due to their proximity to the Sun, these could have been hot enough not to contain any volatiles. After the completion of accretion, all the volatiles of the biosphere, including atmosphere and oceans, were brought by a cometary bombardment. If Earth was formed from planetesimals, which were accumulations of fine dust particles, then the original dust particles would have been at a temperature of about

Figure 5.19: Stardust mission. Test of landing the capsule. Credit: NASA Stardust Mission

1000 K. All carbon, nitrogen and water remained in the gas phase. No volatiles were left on the minor objects. Thus it is hard to understand how oceans or an atmosphere was formed.

Another scenario is that giant planets were formed at the outer reaches of the Solar System. At these distances from the Sun, the solar nebula was cold and minor bodies could be formed by accumulation of frosty particles. They became cometary nuclei. The giant planets grew in mass and reaching 10 to 20 Earth masses, the orbits of these minor bodies were strongly perturbed and they passed to the inner Solar System. These icy objects brought the water reservoir and parts of the atmosphere to the inner planets ([30]).

Comets could also have deposited water to all terrestrial planets including the Moon. The lunar samples from the Apollo programme were found to be devoid of indigenous water. This is a further proof that the Moon was formed during a giant impact event. Later water was found in apatite from mare and highland rocks. Therefore water must have been present during the formation of the mare. A significant delivery of cometary water to Earth and Moon must have occurred shortly after the Moon-forming impact ([36]).

5.5.2   Organic Matter in Comets

The NASA Stardust mission has returned samples form the comet 81P/Wild 2. In January 2004 the spacecraft flew through the comet dust and captured materials (Fig. 5.19, 5.20). The return capsule parachuted back on Earth on Jan. 15, 2006 after a seven year mission.

The primary goal of that mission was to collect samples of a comet and return them to Earth for laboratory analysis. It was first assumed that the comet samples returned by the mission would be composed of tiny grains that formed around other stars, giving the designation stardust. Comets formed in the colder outer parts of the Solar System. The stardust mission showed that these particles form only a minor component.

Silicate crystals were found. This was a mystery because the material to form the Solar System and comets consisted of interstellar dust composed of glassy materials with no crystalline structure. The crystals detected in comets could have been formed out from such interstellar grains those were transformed to crystals by warming in the vicinity of a star. These findings suggest materials from the inner regions of the Solar System could have

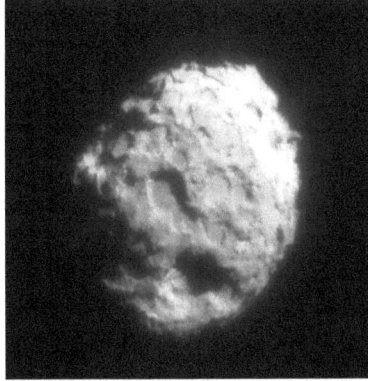

Figure 5.20: Stardust mission. Image of comet Wild. The capsule crossed the halo of the comet and sampled material. Credit: NASA Stardust Mission

traveled to the outer reaches where comets formed.

The organics found in the samples are more primitive than those seen in meteorites. Several samples contain PAHs (polycyclic aromatic hydrocarbons). There are cases of outgassed comets that are known as asteroids, so there seems to be no strict distinction between these two classes of objects.

5.6   Organic Matter in the Presolar Nebula

It is generally accepted that the Sun originated from a collapse of a presolar nebula. Such a nebula becomes gravitationally unstable when gravity is larger than the pressure exerted by the random motion of the individual particles (Jeans criterion).

By observing disks around other stars we can infer the conditions in the presolar nebula and the disk where the Sun and planets formed.

5.6.1   T Tauri Stars

T Tauri stars are pre-main sequence stars. They belong to spectral classes F, G, K, M and their masses are $< 2\,\mathrm{M_\odot}$. Whereas stars on the main sequence in the HRD have reached their equilibrium (see chapter on hydrostatic equilibrium), they are still in a phase of contraction. Their energy supply does not come from thermonuclear reactions but from a release of gravitational energy. The contraction phase to the main sequence depends on the mass of the star but typically lasts for about 100 million years.

T Tauri stars are very active, and they rotate with a period between one and twelve days. They show extremely strong flare emissions (in the X-ray and Radio they are about 1000 times more active than the present Sun). Further characteristics of of T Tauri stars are strong stellar winds (an example is given in 5.21).

Their spectra show higher lithium abundance than the Sun. Lithium is destroyed at temperatures above 2,500,000 K which occurs in the deeper layers of the Sun. Lithium burning is another energy source:

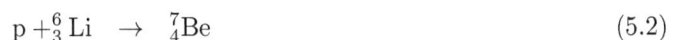

$$\mathrm{p} + {}^{6}_{3}\mathrm{Li} \;\;\rightarrow\;\; {}^{7}_{4}\mathrm{Be} \tag{5.2}$$

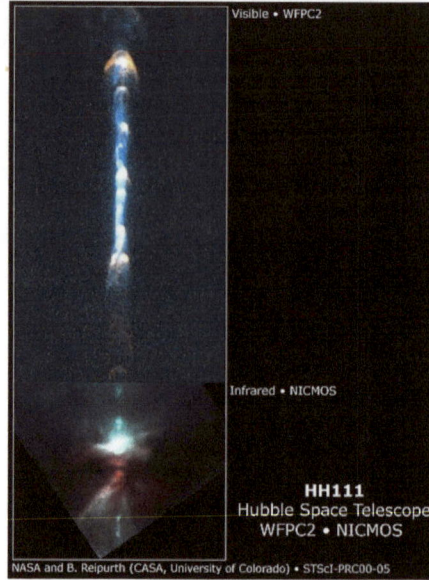

Figure 5.21: This composite image, made with two cameras aboard NASA's Hubble Space Telescope, shows a pair of 12 light-year-long jets of gas blasted into space from a young system of three stars. The jet is seen in visible light, and its dusty disk and stars are seen in infrared light. Credit: NASA and B. Reipurth (CASA, Univ. of Colorado)

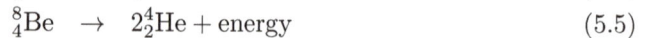

$$^{7}_{4}\text{Be} + \text{e}^{-} \quad \rightarrow \quad ^{7}_{3}\text{Li} \tag{5.3}$$

$$\text{p} + ^{7}_{3}\text{Li} \quad \rightarrow \quad ^{8}_{4}\text{Be} \tag{5.4}$$

$$^{8}_{4}\text{Be} \quad \rightarrow \quad 2^{4}_{2}\text{He} + \text{energy} \tag{5.5}$$

Half of T Tauri stars are surrounded by circumstellar disks. These are progenitors of planetary systems.

Our Sun also was in a T Tauri phase during its early evolution. The rapid rotation of these objects is slowed down by a transfer of angular momentum to the surrounding disk. The mass loss of these stars can be detected spectroscopically by P-Cygni lines. Many spectral lines have a broad emission peak and superimposed to that is an absorption at the blue shifted (short wavelength) edge of the line. Some objects also show inverse P-Cygni line profiles, *i.e.* a redshifted absorption. This indicates an inflow of matter.

Mass loss or mass accretion on a star can be detected by:

- P-Cygni profiles: violet absorption $\rightarrow$ mass loss

- inverse P-Cygni profiles: redshift absorption $\rightarrow$ mass accretion.

Typical mass loss rates are between 10^{-9} to $10^{-7}\,\text{M}_\odot/\text{yr}$ and the stellar wind velocity is about 100 km/s.

The star T Tauri is the prototype of these stars.

5.6.2 Disks around T Tauri Stars, Organic Molecules

With the Spitzer Infrared Spectrograph, molecular emission lines can be studied. Rotational transitions of H_2O and OH as well as non vibrational bands of simple organic molecules such as CO_2, HCN, C_2H_2 are common in the disks of Tauri stars. The gas temperatures in such disks are from 200 to 800 K, with the inner planet formation region being $< 2\,$AU. OH is present in many T Tauri objects and may originate from photodissociation via FUV of H_2O. A comparison of derived HCN-to-H_2O column density ratio to comets, hot cores, and outer T Tauri star disks suggests that the inner disks are chemically active ([16]).

High resolution spectroscopy revealed features of H_2O, OH, HCN, and C_2H_2 as well as two other molecules thought to be abundant in the inner disk, CH_4 and NH_3 ([70]).

5.6.3 Planets in Protoplanetary Disks

Planets form in protoplanetary accretion disks around young stars however, the detection of planets in such disks is difficult. The stars are very active as it has been mentioned above. Therefore, the radial velocity method cannot be applied. The radial velocity method was discussed in the chapter about extrasolar planet detection.

The study of the near infrared emission spectra could provide another method to find such Jovian planets. A massive planet may induce an observable effect on spectral lines that emerge in the disks atmosphere. The spectral signatures of such a planet are:

- permanent line profile asymmetry

- short timescale variability associated with the orbital period of the planet. This comes from the local dynamical perturbation by the orbiting giant planet.

Further details can be found in [83].

Activities

Consider the different bodies in the Solar System and make a ranking concerning the possibility to find life there:

- planets,

- satellites of planets,

- comets

CHAPTER 6

Search for Extrasolar Planets

Abstract: In this chapter we review the different methods for detection of extrasolar planets. The main problem to find these objects comes form

- planets are smaller than stars,
- they only reflect light from a star,
- seen from a great distance they appear very close to their parent star,
- they mainly radiate in the IR.

The main detection methods are velocity measurements, astrometric measurements, direct imaging, microlensing. Satellite missions such as KEPLER, COROT and GAIA are mentioned. Today, it became even possible to detect exoplanetary atmospheres. Finally, the different types of detected exoplanets are reviewed.

The main message of this chapter is that we are at the beginning to understand under which conditions planets form around stars and how many of these planets could be Earthlike.

Keywords: Exoplanets; Detection of Exoplanets; radial velocity method; Doppler method; microlensing; COROT; KEPLER, GAIA; Exoplanets: atmospheres; Exoplanets: types; spectral lines

6.1   Some Basic Considerations

6.1.1   Albedo and Climate

The Albedo of a planet is the diffuse reflectivity of a surface. It is defined as the ratio of reflected radiation from the surface to incident radiation upon it. It depends on the wavelength. The average albedo of a planet depends on its surface and cloud coverage. The average global albedo of the Earth is 30% to 35%. Fresh snow has an albedo of 0.9, charcoal about 0.04. The Earth's ocean has a low albedo; green grass has an albedo of 0.25, desert sand 0.40. The relatively high albedo of the Earth comes from the contribution of clouds (see also Fig. 6.1).

The albedo is extremely important for the weather and climate on a planet. If Earth's climate is colder and there is more snow and ice on the planet, more solar radiation is reflected back out to space and the climate gets even cooler. On the other hand, when warming causes snow and ice to melt, darker colored Earth surface and ocean are exposed and less solar energy is reflected out to space causing even more warming. This is known as the ice-albedo feedback.

Clouds also have an important effect on albedo. They have a high albedo and reflect a large amount of solar energy out to space. Different types of clouds reflect different amounts

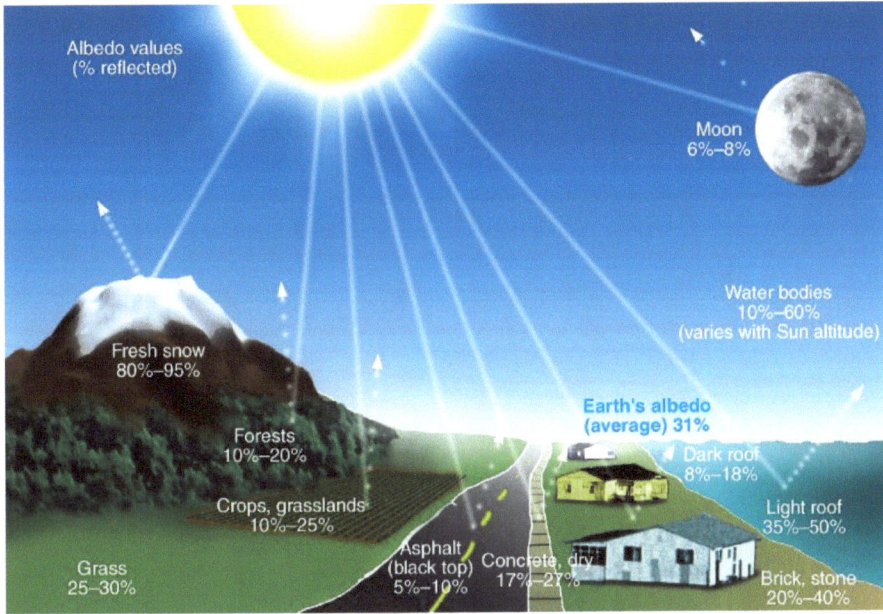

Figure 6.1: Different features on the surface of Earth and the corresponding albedo. Credit: http://www.sci.uidaho.edu/scripter/geog100/lect/03-atmos-energy-global-temps/03-06-albedo-values.htm

of solar energy. If there were no clouds, Earth's average albedo would have dropped by half.

In the Solar System there are two known objects with a very high albedo: Enceladus (moon of Saturn) has an albedo of 0.99, while Eris (dwarf planet) has an albedo of 0.96. Small objects in the Solar System have a low albedo (about 0.05), the nucleus of a comet is also extremely dark with an albedo of 0.04. Such a low albedo is caused by a coverage of organic compounds.

Generally, in Astrophysics there are two types of Albedos:

- The geometric albedo of an astronomical body is the ratio of its actual brightness at zero phase angle (*i.e.*, as seen from the light source) to that of an idealized flat, fully reflecting, diffusively scattering disk with the same cross-section.

- Bond albedo: measures the total proportion of electromagnetic energy reflected.

The Bond albedo is the ratio of the total radiation reflected or scattered by the object to the total incident light from the host star integrated over all wavelengths.

The emitted flux of a star is given by:

$$F_\nu = \Omega_\nu B_\nu(T) \tag{6.1}$$

Where Ω_ν is the solid angle into which the radiation is emitted, and B_ν is the source brightness. The radiation emitted from a black body is given by Planck's law:

$$B_\nu(T) = \frac{2h\nu^3}{c^2} \frac{1}{\exp^{h\nu/(kT)} - 1} \tag{6.2}$$

Table 6.1: Bond Albedo and Geometric Albedo of Solar System Objects.

Name	Bond Albedo	Geometric Albedo
Mercury	0.119	0.138
Venus	0.90	0.67
Earth	0.31	0.37
Moon	0.12	0.11
Mars	0.25	0.15
Jupiter	0.34	0.52
Saturn	0.34	0.47
Enceladus	0.99	1.4
Uranus	0.30	0.51
Neptune	0.29	0.41
Pluto	0.4	0.44–0.61

integrating over all frequencies we obtain Stefan's law

$$F = \int F_\nu d\nu = \pi \int B_\nu(T)d\nu = \sigma T^4 \tag{6.3}$$

Let us consider a surface element dA, $\theta = 0$ is the normal to the planet's surface and θ the angle of the reflected light. The phase integral q_{PH} contains the phase (the angle seen from a planet between the direction planet-Sun and planet-Earth):

$$q_{\mathrm{PH}} = 2\int_0^\pi \frac{F(\Phi)}{F(\Phi = 0)} \sin \Phi d\Phi \tag{6.4}$$

This shows the relation between the Bond Albedo, A_b and the geometric albedo, A_0:

$$A_b = A_0 p_{\mathrm{PH}} \tag{6.5}$$

The solar constant is defined as the solar flux at 1 AU:

$$F_\odot = \frac{L_\odot}{4\pi r_{\mathrm{AU}}^2} = 1.37 \times 10^6 \, \mathrm{erg\,cm^{-2}\,s^{-1}} \tag{6.6}$$

The geometric albedo may by greater or smaller than the Bond albedo depending on surface and atmospheric properties of the object. In Table 6.1 some values for Solar System objects have been mentioned.

6.1.2   Extrasolar Planets: Phase Curves and Albedos

The phase angle is the angle between the light incident onto an observed object and the light reflected from the object. A phase curve describes the brightness of a reflecting body as a function of its phase angle (Fig. 6.2). The phase curve of Mercury is very steep being characteristic of a body on which bare regolith (soil) is exposed to view. At phase angles exceeding 90^0 (crescent phase) the brightness falls off especially in a sharp manner. The

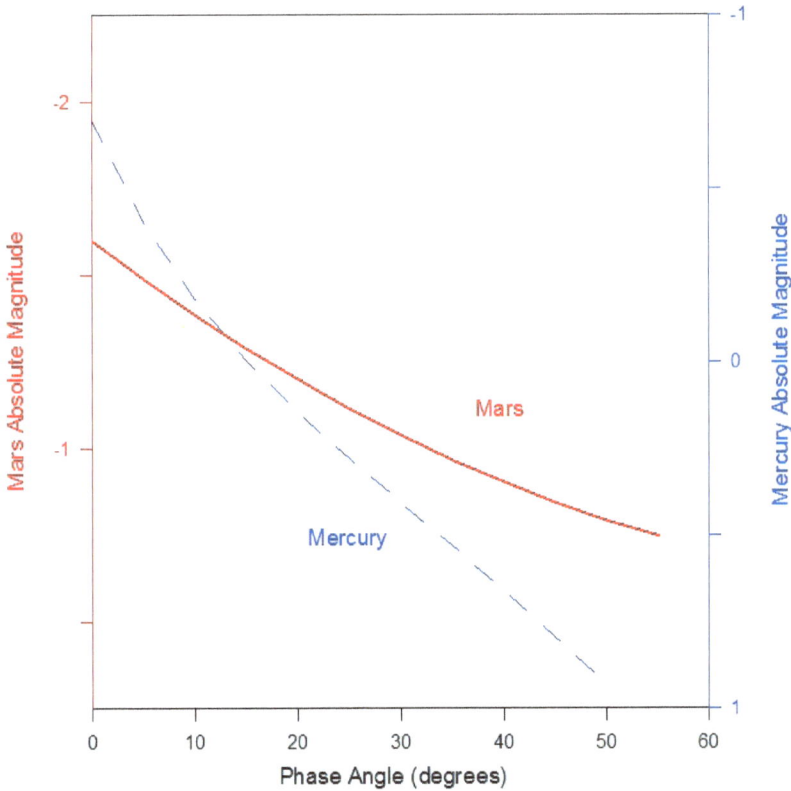

Figure 6.2: Phase curves of Mars and Mercury.

relatively flat phase curve of Venus is characteristic of a cloudy planet[14]. In contrast to Mercury where the curve is strongly peaked approaching phase angle zero (full phase) that of Venus is rounded. Therefore, the phase curve gives information on the surface of a planet.

Telescopic observations of exoplanets reveal them as point sources. This is the same situation as it was for the observation of asteroids in the Solar System before the space missions. Information about their shape and surface composition was obtained by analyzing their light curves (due to their rotation and their irregular shapes the brightness changes) and phase curves.

The phase curves of terrestrial planets strongly differ between

- airless Mercury,

- cloud-covered Venus, and the

- intermediate case of Mars.

The parameters of these different phase functions can be used to characterize exoplanets.

Theoretical considerations about exoplanets and their phase curves and how to determine these were done by[14] and [99].

6.1.3 Observational Aspects

Let us assume that an observer at a distance of 3000 Lyr wants to observe the Solar System. To this observer, the distance Sun-Earth would be only 0.001 arcsec. We can estimate how big a telescope should be used in order to measure this small angular distance: the resolving power of a telescope depends on its diameter D and the wavelength of observation:

$$\alpha = 206265 \times 1.22 \frac{\lambda}{D} \tag{6.7}$$

Thus the telescopic diameter should be 125.8 m (in the visible wavlength). The contrast between the planet's radiation and the star's radiation decreases to the infrared but the resolving power of telescopes also decreases at longer wavelengths.

To resolve the diameter of the Sun at a distance of 3000 Lyr, the aperture should be in the range of 20 km. At present, such instruments seem unrealistic but from space, using interferometry, such instruments could be feasible in the near future. With such instruments we would be able to resolve such details and to observe directly Earth-like planets in our galactic neighborhood. The value of 3000 Lyr $\approx$ 1 kpc was taken because, as it was shown earlier, there is a habitable zone in our Galaxy about that extension and in a zone of about 3000 Lyr extension we can expect to find most candidates for life.

As it has been shown by the above examples, the direct observation of extra- solar planets is extremely difficult, especially when Earth-like extrasolar planets should be detected. However, some indirect methods have already led to reliable results and future space missions being already under realization will ghelp to explore these possible worlds within the next 10 years.

6.2 Detection of Extrasolar Planets

6.2.1 The Two Body Problem

Let's consider two bodies with masses m_1 and m_2, *e.g.* m_1 is the mass of the host star, m_2 the mass of a planet orbiting the host star. The position vectors of the two masses are $\mathbf{r_1}$ and $\mathbf{r_2}$. The force between the two bodies is given by Newton's universal law of gravity: a second body of mass m_2 at position $\mathbf{r_2}$ exerts an attractive force on the first body given by:

$$\mathbf{F}_{12} = -\frac{Gm_1m_2}{r^2}\hat{\mathbf{r}} \tag{6.8}$$

Here $\mathbf{r} = \mathbf{r_1} - \mathbf{r_2}$, G the gravitational constant, and $\hat{\mathbf{r}} = \mathbf{r}/r$. The equation of the relative motion of two mutually graviting bodies can be derived from Newton's law to be:

$$\mu_r \frac{d^2\mathbf{r}}{dt^2} = -\frac{G\mu_r M}{r^2}\hat{\mathbf{r}} \tag{6.9}$$

$$\mu_r = \frac{m_1m_2}{m_1 + m_2} \tag{6.10}$$

$$M = m_1 + m_2 \tag{6.11}$$

The relative motion is equivalent to that of a particle of reduced mass μ_r orbiting a fixed central mass M.

From this equation of the two body problem Kepler's laws can be derived:

- The two bodies move along elliptical orbits; one focus of each ellipse located at the center of mass.

$$\mathbf{r}_{cm} = (m_1\mathbf{r_1} + m_2\mathbf{r_2})/M \qquad (6.12)$$

- A line connecting two bodies sweeps out area at a constant rate, being a consequence of the conservation of angular momentum **L**:

$$\frac{d\mathbf{L}}{dt} = 0 \qquad (6.13)$$

$$\mathbf{L} = \mathbf{r} \times \mathbf{p} = \mathbf{r} \times m\mathbf{v} \qquad (6.14)$$

- The orbital period of a pair of bodies about their mutual center of mass is given by:

$$P_{\mathrm{orb}}^2 = \frac{4\pi^2 a^3}{G(m_1 + m_2)} \qquad (6.15)$$

A planet orbiting a host star can be fully described as a two body system and that in fact both objects move around the barycenter which in most cases lies inside the star because of its mass m_1 being larger than the planet's mass m_2.

Consider the Sun and Jupiter placed at the Earth's orbit. The mass of the Sun is 1000 times that of Jupiter. Is the center of mass of this system still inside the Sun?

6.2.2 Astrometry

Consider a two body problem as was described above. If there is a massive planet (or even more planets) around a star, the star will also move about the center of gravity.

The more the massive one of the two components, the more the center of mass will be located near to that mass. Let us consider the center of mass of the system Sun and Jupiter (the most massive planet in the solar system). Since the mass of the Sun is about 1000 times the mass of Jupiter, the center of mass of such a two body system must be located 1000 times nearer to the Sun than to Jupiter[1]. Therefore, precise position measurements are required to detect the motion of a star around the center of gravity in case of the presence of other planets.

The motion of the barycenter of the Solar System relative to the Sun is shown in Fig. 6.3.

The reflex motion of the host star due to the presence of a planet is in many cases the only means to derive accurate planetary and stellar masses.

The radial velocity method described in the next section yields true masses only in the case when transit observations are possible, *i.e.* when the system is seen nearly edged on.

The astrometric signal can be estimated from

$$\Theta[\mathrm{max}] = 1.91 \times \frac{\mathrm{a}_{\mathrm{comp}}}{[\mathrm{AU}]} \times \frac{[\mathrm{pc}]}{\mathrm{d}} \times \frac{\mathrm{M}_{\mathrm{comp}}}{[\mathrm{M}_{\mathrm{Jup}}]} \times \frac{[\mathrm{M}_{\odot}]}{\mathrm{M}_*} \qquad (6.16)$$

---

[1]In fact it is located just outside the Sun's sphere

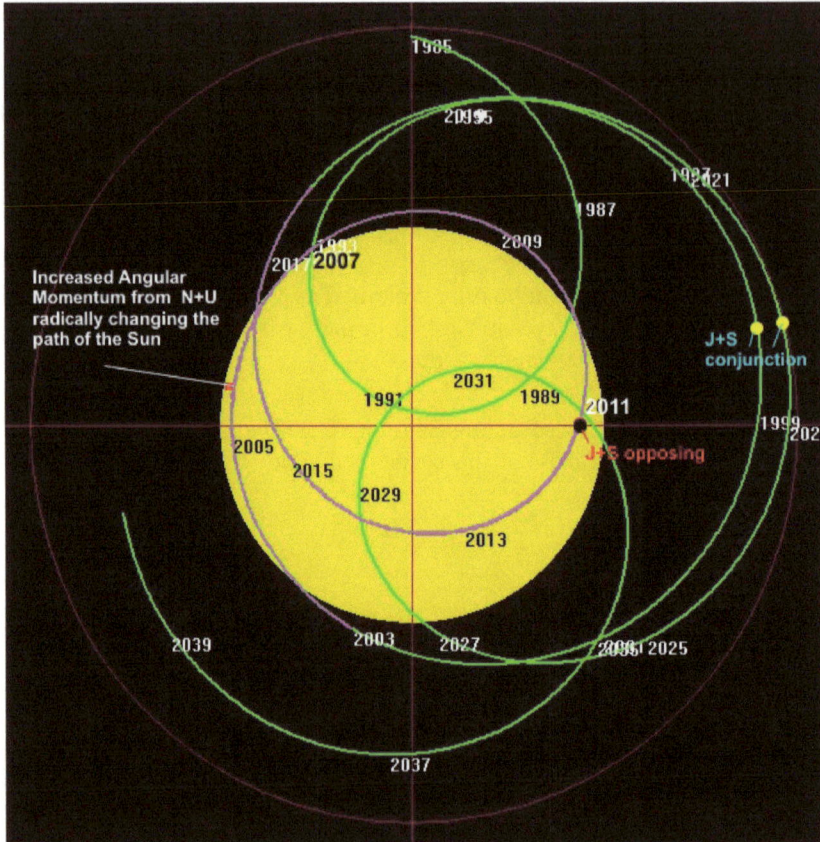

Figure 6.3: Motion of Barycenter of the Solar System relative to the Sun. Credit: Wikimedia.

a_{comp} is the semi-major axis, M_{comp}, the mass of the planetary companion orbiting a star with mass M_* at a distance d.

Let us consider an exoplanet with a mass of 1 M_{Jup} around a solar-mass star in a two years orbit at a distance of 10 pc. Then from Kepler's third law:

$$\frac{a^3}{P^2} = \frac{G}{4\pi^2} (m_* + m_{comp}) \tag{6.17}$$

The astrometric signal will be 0.3 mas = $300\mu as$. A planet having a mass of 10 M_{Jup} at distance 100 pc orbiting a star with mass 0.5 $M_\odot$ in one year will induce an astrometric signal of about 300 μas.

These examples show: the precision needed to find exoplanets with astrometry is in the μas–regime.

The astrometric re-detections of exoplanets were all done with the Hubble Space Telescope (HST): Gl 876 b, 55 Cnc and Epsilon Eridani b. For example, the first astrometrically determined mass of an exoplanet Gl 876 b yielded the following values [7]:

semi major axis: 0.25 ±0.06 mas

inclination: $i = 84 \pm 6^0$

parallax: 214.6 ±0.2 mas

mass of the primary star: $M_* = 0.32 M_\odot$

mass of the planet Gl 876b $= 1.89 \pm 0.34 \, M_{Jup}$

For the planet orbiting the K2V star ϵ Eridani, the following parameters were found [8]:

semi major axis 1.88 ± 0.2 mas

inclination 30.1 ± 3.8^0

$M_a st = 0.83 M \odot$

$M_{Comp} = 1.55 M_J \pm 0.24 M_J$

Cool nearby M dwarfs will be investigated with precision astrometry by the GAIA mission[2].

Another technique applied will be phase-referenced interferometric astrometry. PRIMA (Phase-referenced Imaging and Micro-arcsecond Astrometry) will measure the astrometric wobble of a candidate star due to an exoplanet relative to a close-by 'calibrator' star located within the instrument's observing field (1 arcmin in the PRIMA case) ([9]). The PRIMA is installed at ESO VLTI (Paranal Observatory) designed to enable simultaneous interferometric observations of two objects each with size of at most 2 arsec that are separated by up to 1 arcmin. There are two modes: (i) measure the angular separation between two objects (astrometry mode), (ii) produce images of the fainter of the two objects using a phase reference technique (imaging mode). The PRIMA is managed by ESPRI (Exoplanet Search with PRIma).

6.2.3 Radial Velocity Method

We have seen how astrometric measurements can be used to determine the motion of a star around the center of gravity due to the gravitational influence of the planet. Of course, such a motion also leads to variations in the speed with which the star moves relative to Earth. By precise Doppler measurements we can detect displacements of spectral lines (see Fig. 6.4). Since the star is more massive than the planet, these motions are extremely small and we must be able to measure velocities down to 1 m/s.

---

[2]GAIA (originally an acronym for Global Astrometric Interferometer for Astrophysics) is a European Space Agency (ESA) space mission in astrometry to be launched in August 2013.

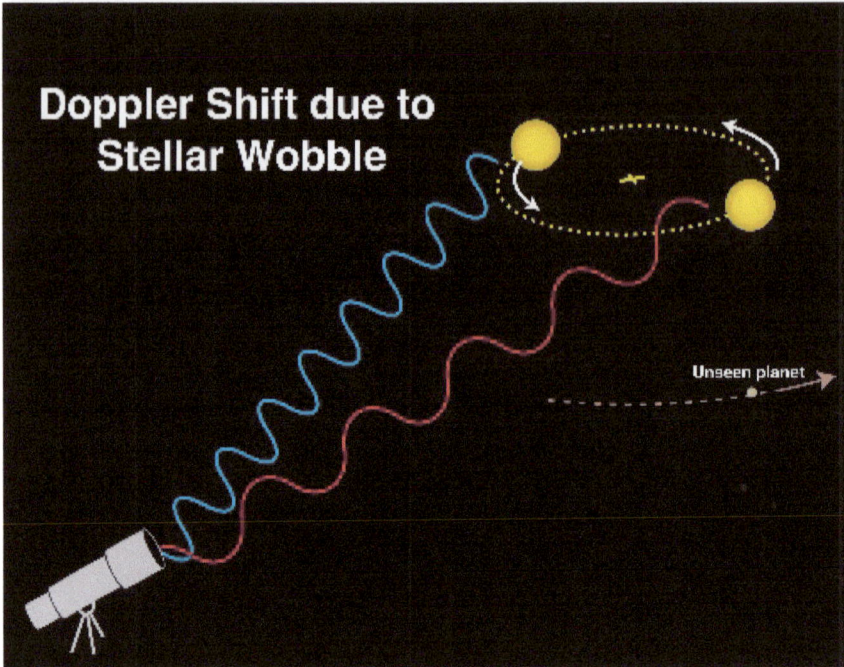

Figure 6.4: The motion of the central star about the barycenter due to the presence of an exoplanet can be measured by a periodic Doppler shift.

The Doppler velocity can be derived from the Doppler formula[3]:

$$\lambda' = \lambda_0(1 + v/c) \qquad (6.18)$$

λ' is the wavelength of the line coming from a source moving with velocity v in radial direction relative to an observer, λ_0 is the velocity of the spectral line when the source is at rest relative to an observer and c is the speed of light. This formula is valid only for non relativistic speeds, *i.e.* $v << c$.

Consider a wavelength of $\lambda = 500$ nm. What would be the amount of Dopplershift when a host star moves at 1 m/s in radial direction (i) toward, (ii) away from the observer?.

Such Doppler shifts are measured with spectrometers. Early spectroscopes were made of prisms, but modern spectroscopes use a diffraction grating. Two examples of modern high resolution spectrometers are the HARPS (High Accuracy radial velocity Planet Searcher) at the ESO 3.6 m telescope in La Silla Observatory and the HIRES at the Keck Telescope.

The method is very precise but requires a high signal-to-noise ratio and therefore works only for objects nearer than 160 ly. Planets being massive and close to their host stars can be easily found (because of the short period of revolution about the host star), however the detection of massive planets at greater distances requires years of high precision observation; for example consider Jupiter. The duration of its orbital period is about 12 years. Planets

---

[3]Ch. Doppler, 1803-1853

with orbits highly inclined to the line of sight from Earth produce smaller wobbles and are more difficult to detect.

The semi-amplitude, K, of the radial velocity of a star of mass M_* that is induced by an orbiting planet of mass M_P is:

$$K = \left(\frac{2\pi G}{P_{\text{orb}}}\right)^{1/3} \times \frac{M_P \sin i}{(M_* + M_P)^{2/3}} \frac{1}{\sqrt{1 - e^2}} \tag{6.19}$$

P_{orb} is the orbital period, i the angle between the normal to the orbital plane and the line of sight, e the eccentricity of the planet's orbit. We see that for $i = 0$ there is no amplitude since we cannot measure any radial velocity component.

The first planet discovered using the technique of radial velocity measurements was 51 Pegasi b ([72]). The companion lies only about eight million kilometres from the star, which would be well inside the orbit of Mercury in our Solar System. This object might be a gas-giant planet been migrated to this location through orbital evolution.

6.2.4 Transit Method

If the Earth lies in or near the orbital plane of an extrasolar planet, the planet passes in front of the disk of its star. Precise photometry reveals such transits. The light variations caused by a transit of a planet show a square-well shape (Fig. 6.5) not dependent on wavelength. The duration of a transit can be estimated from

$$\tau_{\text{tr}} = 13 d_* \sqrt{a/M_*} \sim 13\sqrt{a} \quad \text{[hrs]} \tag{6.20}$$

a is the semi major axis of the planet's orbit, M_* the mass of the star and d_* the stellar diameter in solar diameters. The duration alone does not give any information about the physical nature of the planet. The size of a planet follows from the transit depth because the fractional change in brightness (transit depth) is equal to the ratio of the planet's area to the star's area. As has been pointed out earlier about the common properties of stars, the spectral type of a star tells us its size (for main sequence stars). As an example we give the transit properties of Solar System objects in Table 6.2 from http://kepler.nasa.gov/sci/basis/character.html. P is the orbital period in years, a the semi major axis in AU, T the transit duration in hours, D the transit depth in %, Prop the geometric probability in % and Incl the inclination invariant plane in degrees.

Greater precision of photometric observations is achievable above the Earth atmosphere from satellite missions, and will help to detect planets as small as Earth. One advantage of this method is that larger planets detected from photometric lightcurves can be tested and verified with the radial velocity method since this method works best under $i = 90^0$. The precision of the measurements must be high. For example in the case of HD 209458 the drop in the lightcurve is only 1.7 %.

Giant exoplanets seen from Earth display phases like the Moon. When they transit the host star, the situation becomes similar to New Moon, shortly before vanishing behind the host star or after reappearing from it, they become nearly fully illuminated, which resembles Full Moon. This causes an additional small light variation of the system.

However, there are also other sources of light variation of a star therefore one must ve careful in the interpretation of light curve variations:

- starspots,

Table 6.2: Transit Properties of Solar System Planets.

Planet	P	a	T (h)	D (%)	Prop.	Incl
Mercury	0.24	0.39	8.1	0.0012	1.19	6.33
Venus	0.62	0.72	11.0	0.0076	0.65	2.16
Earth	1.0	1.0	13.0	0.0084	0.47	1.65
Mars	1.88	1.52	16.0	0.0024	0.31	1.71
Jupiter	11.86	5.2	29.6	1.01	0.089	0.39
Saturn	29.5	9.5	40.1	0.75	0.049	0.87
Uranus	84.0	19.2	57.0	0.135	0.024	1.09
Neptune	164.8	30.1	71.3	0.127	0.015	0.72

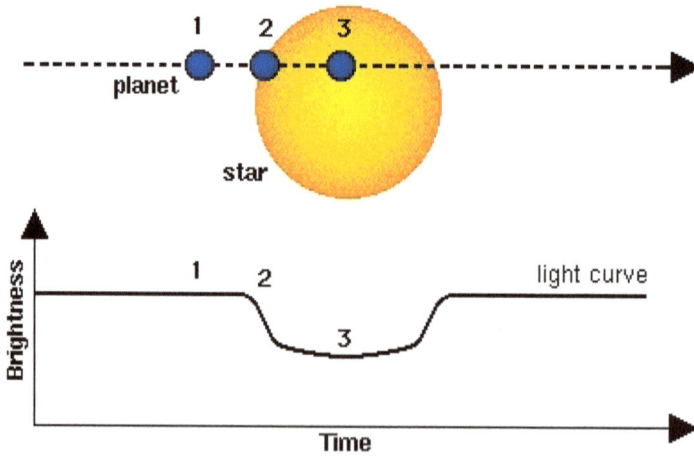

Figure 6.5: The transit of a planet can be observed by a dip in the lightcurve of the star.

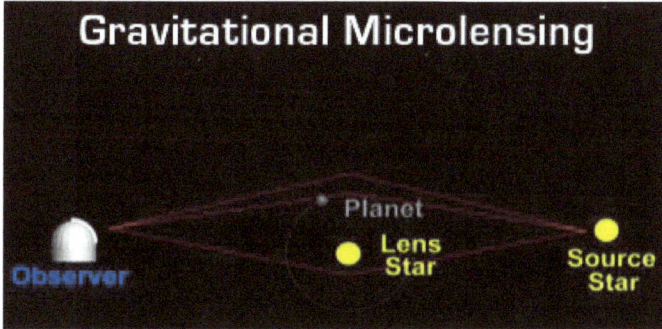

Figure 6.6: Principle of microlensing.

- intrinsic stellar variability.

Starspots are rotationally modulated.

Two important satellite missions for the detection of exoplanets are the KEPLER mission and the COROT mission. The KEPLER mission uses about 100 000 stars that are scanned almost continuously.

6.2.5   Microlensing

Gravitational microlensing (Fig. 6.6) occurs when the gravitational field of a star acts like a lens, magnifying the light of a distant background star. If the foreground lensing star has a planet, then that planet's own gravitational field can make a detectable contribution to the lensing effect. This can be observed in the variation of the lightcurve as a secondary bump (see Fig. 4.12).

D_L denotes the distance between the observer O and the lens, D_{LS} is the distance between the lensing star and the source S and D_S is the distance between the observer and the source. An amplification of the light due to microlensing can be observed when the object moves within the Einstein radius defined by:

$$\Theta_E = \sqrt{\frac{4GM}{c^2}\frac{D_{LS}}{D_L D_S}} \tag{6.21}$$

Under favorable circumstances, planets as small as Earth would be detected.

The main disadvantage of this method is that lensing cannot be repeated since the alignment never occurs again. The detected planets will be several kpc away, so making the other methods not applicable to test the observation. On the other hand, by a systematic observation we would be able to get the basic information about how common are earthlike planets in the Galaxy.

The observations will be made by robotic telescopes distributed worldwide. One network is NASA/NSF-funded OGLE, one other project will be PLANET (Probing lensing Anomalies NETwork).

The planet OGLE-2005-BLG-390Lb is known as a super Earth planet being located at a distance of 21,500 Ly from the Solar System. The host star has spectral type M4, the mass of the planet is 5.5 Earth masses and its distance to the host star is about 2.6 AU. Since the host star is a cool M star, the surface temperature expected on the planet is only 50 K.

It was detected by the Danish 1.54-m telescope at ESO, as the telescope being the part of the PLANET network.

6.2.6 Direct Observations

Direct observation means imaging. As it has already been stated that extrasolar planets are very faint objects located near much brighter stars, thus, are extremely difficult to observe directly. The light from the star and the planet has to be separated. The first planet discovered by this technique orbits a brown dwarf. In 2008, three new planetary systems orbiting A-type main sequence stars were announced to be detected. The star HR8799 seems to contain four directly imaged planets. The planets orbiting this star have wide orbits (larger than 25 AU) and are very young and very massive ([71]).

The star Fomalhaut (αPsa) (see also Fig. **??**) is at a distance of 7.7 pc from Earth its spectrum showing a strong IR excess indicating a circumstellar disk. The inner edge of the toroidal debris disk is found at a distance of 133 AU from the star the belt having an extension of 25 AU. In 2008, a planet just orbiting at the inner edge of the debris disk was found using HST observations. The mass determination is not definite yet but the planet's mass is somewhere between half the mass of Neptune and three times the mass of Jupiter. Observations of the star's dust ring by the Atacama Large Millimeter Array point to the existence of two planets in the system, neither one at the orbital radius proposed for the HST-discovered Fomalhaut b ([11]).

6.2.7 Pulsar Timing

Stars with masses larger than 1.4 solar mass explode into a supernova. The explosion starts with an implosion, but the inner mass contracts, the outer parts are expelled. The final stage of a stellar evolution of massive stars is a neutron star, whereas stars with masses larger than about 4 solar masses end as black holes.

Pulsars are rapidly rotating neutron stars. Stars generate their energy by thermonuclear fusion reactions. These reactions end up with the formation of iron. Therefore, an iron core develops. This iron core collapses to about the size of 10 km. Supernovae eject newly formed massive elements into interstellar space. The mass of the core is 1.4 $M_\odot$. Neutron stars have enormously strong magnetic fields. Electrons and positrons moving in the neutron star's magnetic field produce radiation. This radiation is beamed away from the poles of the neutron star. As the neutron star rotates, these beams sweep around like the beam of a lighthouse. As the beam sweeps past an observer the neutron star appears to pulse on and off. It therefore has been named as pulsar.

The intrinsic rotation of a star is quite regular. Slightest variations in the pulses caused by the pulsar's motion can be easily measured. During orbiting of a planet, the pulsar and the planet move around the center of gravity. These tiny motions can be measured so precisely from the pulses, such that the masses down to 1/10 Earthmasses can be detected. In 1992, A. Wolszczan and D. Frail, used this method to discover planets around the pulsar PSR 1257+12. This was in fact the first confirmation of a detection of extrasolar planets.

[21], report deep imaging observations of the young, nearby star AB Pic. They detected a faint, red source 5.5" south of the star with JHK colors compatible with that of a young substellar L dwarf. Follow-up observations at two additional epochs confirmed that the faint red object is a companion to AB Pic rather than being a stationary background object. A low resolution K-band spectrum indicates an early-L spectral type for the companion. Finally, evolutionary model predictions based on the JHK photometry of AB Pic b indicate a mass of 13 to 14 MJ$_J$ if its age is $\sim$30 Myr.

Planets around pulsars and other evolved stars were discussed in the book of [111]. So far eight circumbinary planets (these are planets orbiting a double star) have been discovered in this way.

6.3   Atmospheres of Transiting Planets

The atmospheres of exoplanets can be studied in three ways.

6.3.1   Transmission Spectroscopy

During the transit the planet passes in front of the star. The amount of dimming is thought to be the direct measure of the relative size of the planet compared to the size of the star. Let's consider a planet with an atmosphere. The stellar light passing through its atmosphere is absorbed by atoms and molecules in the planet's atmosphere. Moreover, the effective size of the planet will appear larger at particular wavelength of high absorption. A fractional increase in transit depth in an absorption line occurs:

$$\frac{\Delta F}{F} = 2\frac{\Delta R_p}{R_*}\frac{R_p}{R_*} \tag{6.22}$$

R_p is the planetary radius, R_* the stellar radius, ΔR_P is the change in radius due to absorption by a molecule. The typical size variation as a function of wavelength is proportional to the atmospheric scale-height:

$$H = \frac{kT}{\mu g} \tag{6.23}$$

g is the planet's surface gravity, T the atmospheric temperature, μ the mean molecular weight of the gas. For the Earth, this scale-height is about 10 km, while for a typical hot Jupiter being a few hundred km. This corresponds to an increase in transit depth of

- $\sim 10^{-7}$ for Earth

- $\sim 10^{-4}$ for a typical hot Jupiter.

Several elements have been identified so far in planetary atmospheres: Sodium, potassium, hydrogen, carbon, oxygen and even water. [104], found that absorption by water vapor is the most likely cause of the wavelength-dependent variations in the effective radius of the planet at the infrared wavelengths 3.6, 5.8 and 8 microns. The larger effective radius observed at visible wavelengths may arise from either stellar variability or the presence of clouds/hazes.

6.3.2   Secondary Eclipse

When the planet passes behind the star, the light from it is blocked by the star and the observer sees only the light of the star. Comparing the flux before, during and after the eclipse, one can get information about the additional contribution of a planet to the light of the system host-star and planet.

The planet reflects light that is proportional to the Albedo, A, and the radius of the planet R_P and the semi major axis a. The change in flux (comming from the star and reflected flux of a planet; wavelength dependence is neglected here) becomes therefore:

$$\frac{\Delta F}{F} = A \left(\frac{R_P}{a}\right)^2 \tag{6.24}$$

Let us consider two examples:

- Earth-sized planet orbiting at 1 AU: $\Delta F/F < 1.8 \times 10^{-9}$

- Hot Jupiter at 0.03 AU $\Delta F/F < 4 \times 10^{-4}$.

The other additional source is the thermal emission of a planet. This can be estimated:

$$\frac{\delta F}{F} = \frac{F_p}{F_*} \left(\frac{R_p}{R_*}\right)^2 \tag{6.25}$$

So the contributions of a planet to the toal flux come from (i) reflected starlight (ii) thermal planetary emission. The thermal planetary emission is observed in the IR.

6.3.3 Phase Curve

Like the Moon and inner planets in the Solar System (Venus and Mercury), extrasolar planets also show phases. When the planet's thermal emission was observed, it was noted that the variations in light are caused by the temperature distribution in the planets atmosphere. If both the day- and night-side have the same temperature no phase variations are seen, while the strongest variations are seen if the day-side re-emits all the absorbed stellar radiation before it can be transported to the planets night-side. Therefore, phase curve measurements of the thermal emission from planets allowing the re-distribution of the absorbed stellar light from the day- to the night-side were determined.

6.4 Satellite Missions and Ground Observations

6.4.1 Ground Based Observations

Ground based observations of exoplanets are limited mainly by our atmosphere. In astronomy the term seeing is used. It refers to the blurring and twinkling of stars you can even see with the naked eye. Since there are turbulent layers in the atmosphere the optical refractive index in the Earth's atmosphere varies.

Seeing is often described by the diameter of the seeing disc being the point spread function for imaging through the atmosphere. The point spread function denotes the best possible angular resolution which can be achieved by a telescope. Under the best seeing conditions, the point spread function is about 0.4 arcsec which can only be achieved in high-altitude observatories such as Mauna Kea or La Palma, or at ESO facilities in Chile.

The effects of astronomical seeing were responsible for the "detection" of the canals on Mars which later turned out to be optical illusions.

Active optics (Fig. 6.7) technology is used with reflecting telescopes. The mirrors of large telescopes are shaped actively to prevent deformation of wave front. Active optics also made the construction of telescopes in the class of 8 m possible. The classical scheme for a mirror telescope is a primary and a secondary mirror, both had to be very thick in order to hold their shape when moving the telescope across the sky. Therefore, the maximum

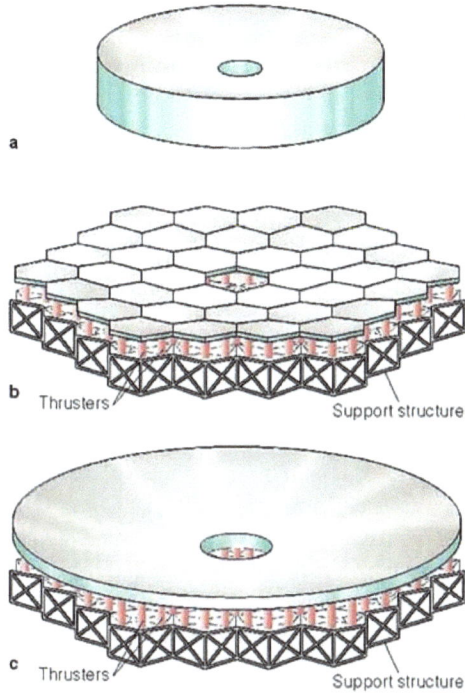

Figure 6.7: Principle of active optics. Each mirror segment is computer controlled and deformable. Large telescope mirrors consist of many cells. Source: http://www22.homepage.villanova.edu/rex.saffer/SESAME/telescopes.html.

diameter of such telescopes was limited to about 5-6 m because of the enormous weight of the primary.

Since 1980s thin mirrors are being used. An array of actuators aligns the mirrors in shape. The large primary mirror of a reflecting telescope was thus segmentated into small mirrors. When these actuators are coupled with a quality-of-image detector, the actuators can be controlled by a computer software in such a way as to restore deformed wavefronts, the deformation coming from the Earth's atmospheric disturbances.

Adaptive optics operates on much shorter timescale to compensate atmospheric effects, rather than mirror deformation.

All the below mentioned (small) telescopes use transit observations (photometry) to detect exoplanets:

- The Trans Atlantic Exoplanet Survey, TrES uses three 10 cm telescopes located at Lowell Observatory, Palomar Observatory and the Canary Islands to locate exoplanets. The telescopes are Schmidt telescopes equipped with CCD cameras and automated search routines. Four planets have been discovered so far, the magnitude [4] of the host stars is between 11 and 12.

- The XO telescope is located at an altitude of 3000 m in Hawaii and consists of a pair of 200 mm telephoto lenses. The magnitudes of the host stars are about 11. The smallest object detected so far is XO-2 with a mass of 0.6 M_J.

- The Hungarian Automated Telescope Network (HATNet) consists of six fully automated telescopes and is also used to find bright variable stars. 29 extrasolar planets have been discovered so far.

- The SuperWASP (Wide Angle Search for Planets) consists of two robotic observatories. SuperWASP-North is at the Roque de los Muchachos Observatory on La Palma in the Canaries, WASP-South at the South African Astronomical Observatory. The observatories consist of an array of eight Canon 200 mm f1.8 lenses and 2k× 2k CCDs. More than 70 extrasolar planets have been detected so far. The limiting magnitude of the host star is 15.

6.4.2 Space Missions

COROT

COROT means Convection Rotation et Transits planetaires (French). This is the mission led jointly by the French Space Agency CNES and the European Space Agency, ESA. The two main objectives of this mission were (i) search for extrasolar planets with short orbital periods and (ii) to measure solar-like oscillations in stars (asteroseismology). It was launched on 27 December 2006. In May 2007, the first extrasolar planet, COROT-1b was detected. Operations are underway and will end by 2013.

The instrument is a 27 cm telescope equipped with four CCD detectors. The CCDs are cooled to -40^{0}C. Stars are monitored for brightness variations due to planetary transits. The magnitude range of the observed stars is from 11 to 16 in the visual. Stars brighter than 11 saturate the CCD detectors, while stars dimmer than 16 become too faint.

---

[4]see chapter 7.1.2 for definition of stellar magnitude

Figure 6.8: The star fields that are observed by the KEPLER mission. Source: NASA.

KEPLER

KEPLER is a NASA mission to discover Earth-like planets orbiting other stars. It was launched in March 2009 and the planned mission lifetime is 3.5 years. Only a small portion of the Milky Way Galaxy has been observed. The main question that needs to be answered with that mission would be how many of the several hundred billion of stars in our Galaxy have Earth-size planets. KEPLER has a high sensitivity photometer that continually monitors the brightness of about 145,000 stars in a fixed field of view. Thus, transiting planets would be detectable.

The camera is made up of 42 CCDs at 2200 × 1024 pixels. The CCDs are read out every six seconds. Due to a preselection of interesting stars, only 5% of the pixels are actually read out and stored. The satellite is never occulted by Earth and its photometer is not influenced by straylight from Earth. The target fields are in the constellations of Lyra, Cygnus and Draco (Fig. 6.8). This is also well out of the ecliptic plane, so there is no danger of sunlight entering the high sensitive CCDs. Another reason for having chosen that field is that it is never obscured by the Solar System's Kuiper belt or by the asteroid belt. The stars observed have about the same distance from the galactic center than our Sun.

As we have seen, the probability that a planetary orbit is along the line-of-sight to star is given by the diameter of the star divided by the diameter of the orbit.

- Earth-like planet at 1 AU: probability 0.465%

- at 0.72 AU (Venus) the probability is 0.65%.

The KEPLER field has a size of 115 deg^2 (the Hubble Space Telescope field has only 10 deg^2). KEPLER must see at least three transits to confirm that the dimming of a star was caused by a transiting planet.

By Feb. 2011, 1235 planetary candidates were found circling 997 stars. Of them 69 planets were Earth-size, 228 Super-Earth-size, 662 Neptune size, 165 were Jupiter-size and 19 were twice the size of Jupiter. 54 planets were within the habitable zone.

To see the rapid progress, on December 5 2011 the KEPLER team announced that they had discovered 2326 planetary candidates:

- 207 similar in size to Earth

- 680 super Earth-size

- 1181 Neptune-size

- 203 Jupiter-size

- 55 larger than Jupiter.

Due to more stringent criteria, the number of planets in habitable zones has decreased to 48.

6.4.3 Future Satellite Missions and Ground Based Observations

Here we mention GAIA. GAIA (originally an acronym for Global Astrometric Interferometer for Astrophysics) is a European Space Agency (ESA) space mission in astrometry to be launched in August 2013. It is a successor to the HIPPARCOS mission. The mission aims to compile a catalogue of approximately 1 billion stars, or roughly 1% of stars in the Milky Way. It will monitor each of its target stars about 70 times to a magnitude 20 over a period of 5 years. Its objectives comprise:

- determining the positions, distances, and annual proper motions of 1 billion stars with an accuracy of about 20 μas (microarcsecond) at 15 mag, and 200 μas at 20 mag

- radial velocity measurements with expected detection of tens of thousands of extrasolar planetary systems

- the other missions aims are not relevant here: detection of distant quasars, three dimensional map of the Milky Way.

GAIA (Fig. 6.9) will be placed at Lagrangian point L2 which lies at a distance of 1.5 million kilometres from the Earth in the anti-Sun direction and co-rotates with the Earth in its 1-year orbit around the Sun. It will be launched in 2013.

Other upcoming missions are DARWIN and TPF. The reader should consult to websites to check whether these missions will be done.

6.5 Types of Extrasolar Planets

The detection of ore than 1000 exoplanets has led to their classification.

Figure 6.9: The GAIA telescope. The payload module is built around the hexagonal optical bench (∼3m diameter) which provides the structural support for the single integrated instrument. Credit: ESA

6.5.1   Hot Jupiters

These include objects that have at least the size and mass of Jupiter.

$$M_{\text{Jupiter}} = 1.89 \times 10^{27}\,\text{kg} \qquad R_{\text{Jupiter}} = 142,984 - 133,708\,\text{km}$$

The main difference to the largest planet in our Solar System is however, that hot Jupiters are found extremely close to their host star. If there would be a hot Jupiter in the Solar System, its orbit would be inside the orbit of Mercury. Further properties are:

- gaseous planets,

- high atmospheric temperatures,

- they are tidally locked which means their rotation rate equals their orbital period,

- unusual hemispheric insolation; there are permanent day and night sides.

Fast winds (sonic) may occur in their atmospheres. The dynamical state of their atmosphere can be measured by the analysis of spectral line profiles (see Fig. 6.11). Lines are broadened well beyond their quantum mechanical width through Doppler shifts associated with the thermal motions of atomic and molecular constituents (Doppler broadening) and through the effect of collisions of these atoms and molecules with other gas constituents (pressure broadening). The relative width of Doppler and pressure broadening depend on the local conditions of density and temperature in the atmosphere.

From the theory of stellar atmospheres, it is quite clear that line profiles can be written as a Voigt profile. The Voigt profile is a superposition of two components.

1. The Lorentz profile of a spectral line is given by

$$L(\nu) = \frac{\gamma}{(2\pi\Delta\nu)^2 + (\gamma/2)^2} \qquad (6.26)$$

$\gamma = 1/\tau$ denotes the damping constant, τ measures the mean lifetime of an excited state and γ is given as radiative damping. This depends on the duration of the excited state and collisional damping which takes into account collisions with other particles. The radiative damping depends on the the lifetime of the energy levels n, m being relevant for the absorption process. Typical lifetimes are in the order of 10^{-8} s so that $\gamma = 1/\tau$.

2. The Doppler broadening results from the motion of the radiating particles which is reflected in the Doppler formula:

$$\frac{\Delta\nu_D}{\nu_0} = \frac{\Delta\lambda_D}{\lambda_0} = \frac{V_0}{c} \tag{6.27}$$

V_0 is the most probable velocity of an absorbing atom that is in a medium at temperature T

$$V_0 = \sqrt{2kT/m} \tag{6.28}$$

Assuming a Maxwellian distribution, the Doppler profile is:

$$D(\nu) = \frac{1}{\sqrt{\pi}\Delta\nu_D} exp[-(\Delta\nu/\Delta\nu_D)^2] \tag{6.29}$$

The Voigt profile is obtained by a convolution of Dopplerprofile and damping. It is given by:

$$\Phi_\nu(\nu - \nu_0) = \frac{1}{\pi^{3/2}} \frac{\Delta\nu_p}{\Delta\nu_D} \int_{-\infty}^{\infty} \frac{1}{(\nu' - \nu_0)^2 + \Delta\nu_p^2} \exp\left[-\frac{(\nu - \nu')^2}{\Delta\nu_D^2}\right] d\nu' \tag{6.30}$$

$\Delta\nu_p$ and $\Delta\nu_D$ measure the pressure- and Doppler broadening, Φ_ν is the line profile function. One can derive:

$$\Delta\nu_p = 0.02 - 0.05 \left(\frac{P}{1 \text{bar}}\right) \left(\frac{T}{1500 \text{K}}\right)^{-1/2} \text{cm}^{-1} \tag{6.31}$$

and

$$\Delta\nu_D = 0.14 \left(\frac{m_{H_2}}{m_{mol}}\right) \left(\frac{T}{1500 \text{ K}}\right)^{1/2} \left(\frac{\lambda_0}{1 \, \mu\text{m}}\right)^{-1} \text{cm}^{-1} \tag{6.32}$$

An example for a hot Jupiter is the object HD 189733b. This exoplanet is a typical hot Jupiter with a semi major axis of 0.0312 ($\pm$ 0.0004)AU. Because it is so close to its parent star, it takes only 2 days for one revolution.

The Jupiter-sized planet, called HD 189733b (Fig. 6.10), is too hot for life. But the Hubble observations are a proof-of-concept demonstration that is the basic chemistry for life can be measured on planets orbiting other stars because CO_2 was detected in its atmosphere.

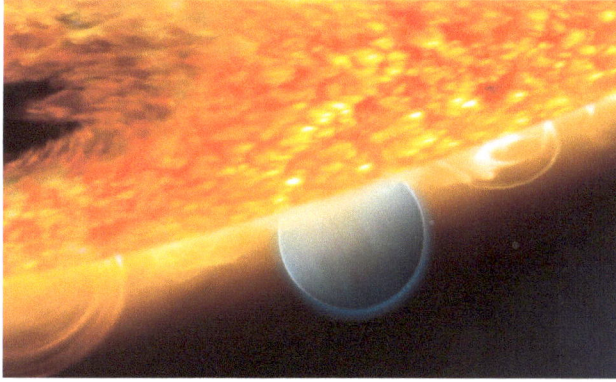

Figure 6.10: Artist's View of Exoplanet Orbiting the Star HD 189733.Credit: ESA, NASA, M. Kornmesser (ESA/Hubble), and STScI

This exoplanet was observed with the Hubble Space Telescope spectrograph. The wavelength investigated was in the range 580.8-638.0 nm with a resolving power of R=5000. Absorption from the NaI doublet within the exoplanet's atmosphere at the 9 sigma confidence level within a 0.5 nm band (absorption depth $0.09 \pm 0.01\%$) was used to measure the doublet's spectral absorption profile. The observations indicate the presence of a high-altitude silicate haze ([44]). IR Spectra were taken with the Spitzer telescope [5].

6.5.2 Hot Neptunes

A hot Neptune is an extrasolar planet at close distance to its parent star, equal to the size and mass of Neptune or Uranus. The first hot Neptune discovered (Aug 2004) was Mu Arae c (or HD 160691c). The discovery was made with the aid of the High Accuracy Radial Velocity Planet Searcher (HARPS) spectrograph (Fig. 6.12), at the European Southern Observatory's La Silla Observatory in Chile. The instrument has been built to obtain very high long term radial velocities at an accuracy of 1 m/s on the order.

The central star HD 160691 is a G3 IV-V object with a mass of about 1.1 solar mass and a radius of 1.36 solar radii. It has an age of about 6.34 ± 0.4 Gyr. The mass of HD 160691c is 14 Earth masses. It is comparable to the object Gliese 436 b. The semi major axis of its orbit is 0.09 AU, the eccentricity is 0.172, the periastron[6] is at 0.075, the apastron[7] at 0.106 AU. The orbital period is only 9.63 days. The planet must be hot because of its closeness to Mu Arae. The discoverers chose for it an albedo of 0.35.

KEPLER has found hundreds of Neptune-size (2-6 $R_\oplus$) planet candidates within 0.5 AU of their stars. There could be two ways to explain these objects:

- core-nucleated accretion and

- outgassing of hydrogen from dissociated ices.

---

[5]Spitzer's telescope is a lightweight reflector of Ritchey-Chrtien design, with a mirror measuring 85 centimeters in diameter. Weighting less than 50 kg, it has been designed to operate at an extremely low temperature to reveal the presence of water. Launch date was Aug 25, 2003

[6]This is the point on the orbit of the planet nearest to its central star

[7]Point at which the distance between the planet and its central star reaches maximum

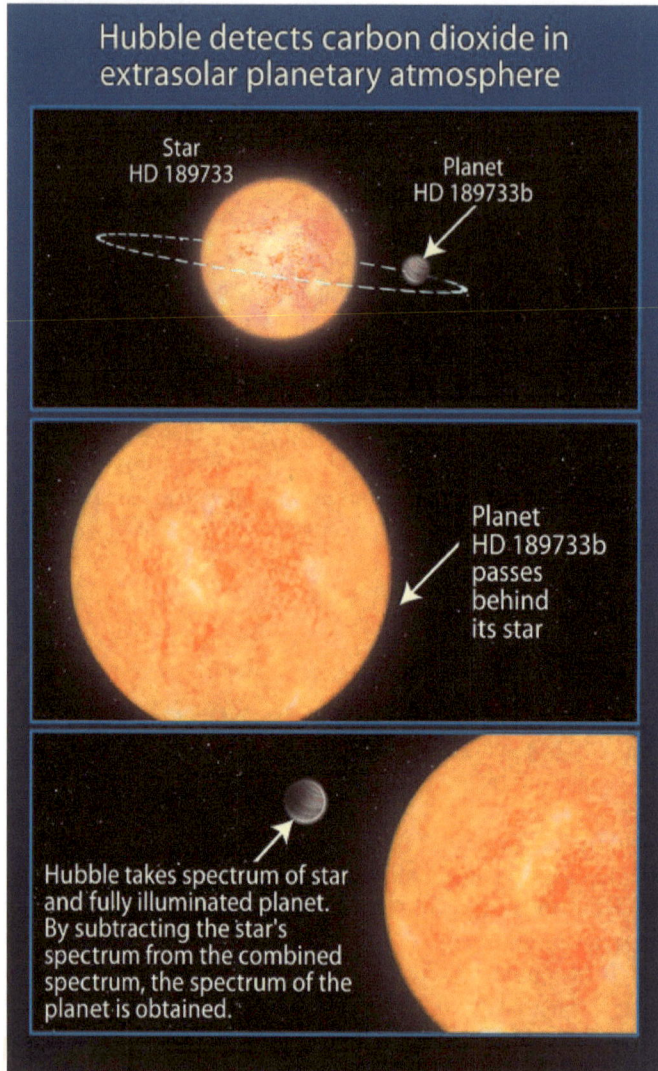

Figure 6.11: Explanation of how to detect spectral signatures on an exoplanet that transits in front of its host star. Credit: NASA press release.

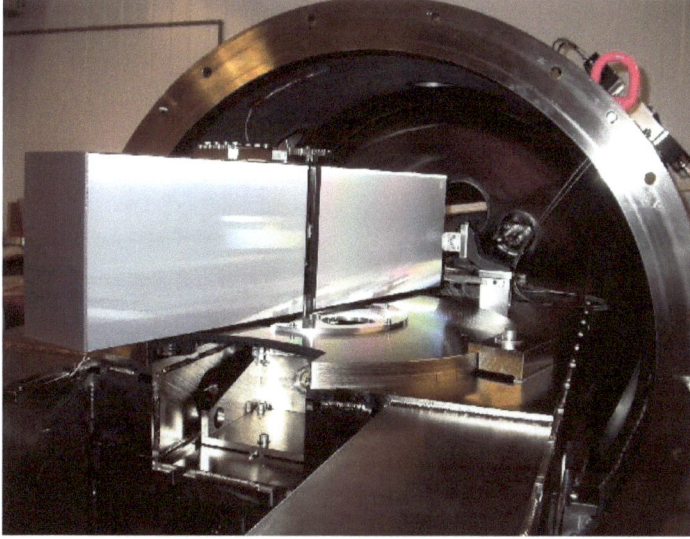

Figure 6.12: The HARPS instrument is an echelle spectrograph kept in a vacuum tank (partly removed here) to avoid spectral drift due to temperature and air pressure variations. Credit: ESO, La Silla, the HARPS team.

Neptune-size planets at $T_e = 500$ K with masses as small as a few times that of Earth can plausibly be formed by core-nucleated accretion coupled with subsequent inward migration (for more details see [87]).

6.5.3   Earthlike Exoplanets

The Gliese Catalogue attempts to list all stars within 25 parsecs (81.5 light-years) of Earth (see also Fig. 6.13). Numbers in the range of 1.0 to 965.0 are from the second edition, Catalogue of nearby stars (1969). The integers represent stars that were in the first edition, while the numbers with a decimal point are used to insert new stars for the second edition without destroying the original order.

The Gliese 581 system contains 5 planets (Fig. 6.14). The central star is a red dwarf at a distance of 20.5 light years,

Gliese 581c is one of the most Earth-like planets discovered to date. It is the third of five planets orbiting Gliese 581 completing each orbit in a mere 13 days. Significantly, it is one of the smallest known exoplanets, measuring only 1.5 times the Earth's diameter and only 5 times its mass, being almost certainly a rock like our own world. It orbits close to the band around its star that is known as the "habitable zone", the only region where conditions are mild enough that water can exist in liquid form. Closer to the star all water will turn to vapor; further away water will turn to ice. But near the habitable zone water can remain liquid and life as we know it could potentially exist.

The fourth planet in the system, Gliese 581d, is even a better candidate for habitability. In 2009 Gliese 581e, was announced with an approximate mass of 1.9 earth masses.

Planets Gl 581c and Gl 581d are near to, but outside, that can be considered as the conservative habitable zone. Planet c receives 30% more energy from its star than Venus from the Sun, with an increased radiative forcing caused by the spectral energy distribution of Gl

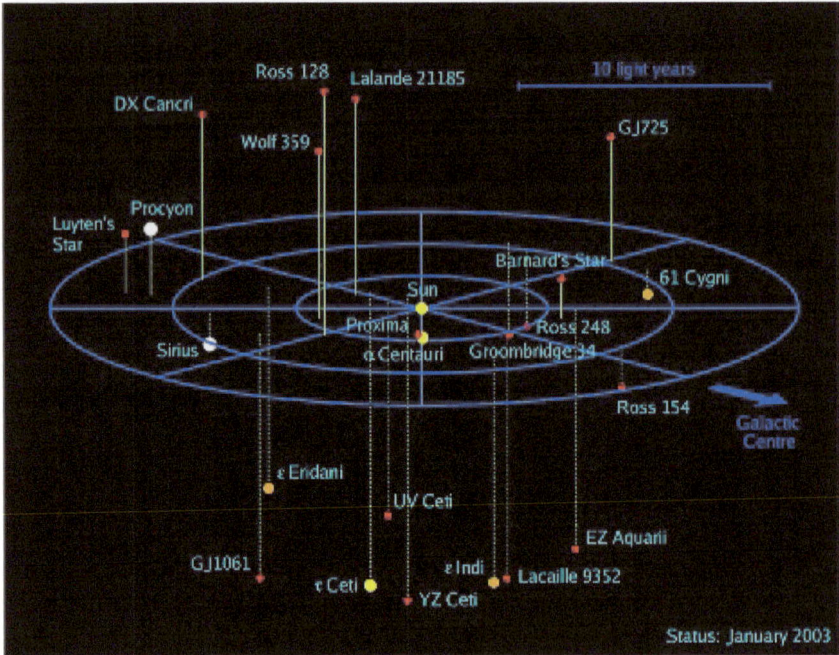

3D Map of Known Stellar Systems in the Solar Neighbourhood

ESO PR Photo 03c/03 (13 January 2003)   ©European Southern Observatory

Figure 6.13: The nearest stars.

The Planetary System in Gliese 581
(Artist's Impression)

ESO Press Photo 22a/07 (25 April 2007)

This image is copyright © ESO. It is released in connection with an ESO press release and may be used by the press on the condition that the source is clearly indicated in the caption.

Figure 6.14: Artist's impression of the Gliese 581 system. Credit: ESO

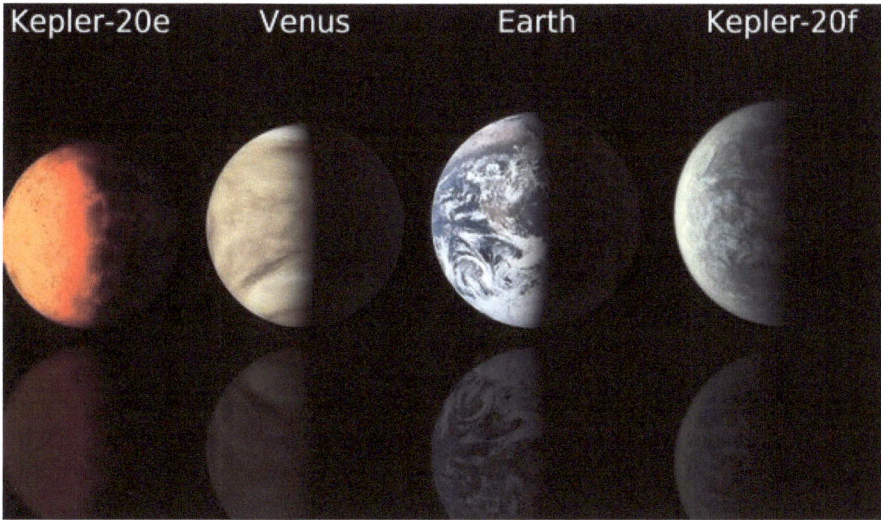

Figure 6.15: The Kepler -20 system.

581. This planet is thus unlikely to host liquid water, although its habitability cannot be positively ruled out by theoretical models due to uncertainties affecting cloud properties and cloud cover. Highly reflective clouds covering at least 75% of the day side of the planet could indeed prevent the water reservoir from getting entirely vaporized. Irradiation conditions of planet d are comparable to those of early Mars, which is known to have hosted surface liquid water. Thanks to the greenhouse effect of CO_2-ice clouds, also invoked to explain the early Martian climate, planet d might be a better candidate for the first exoplanet known to be potentially habitable ([97]).

NASA's KEPLER Mission uses transit photometry to determine the frequency of Earth-size planets in or near the habitable zone of Sun-like stars. Further examples of exoplanets detected by KEPLER are given in Fig. 6.16.

An example of a rocky planet is Kepler-10b which has the following parameters: $M_P = 4.56 \, M_\oplus$, $R_P = 1.416 \, R_\oplus$, and $\rho = 8.8 \, \text{g/cm}^3$ ([5]).

The Kepler-20 system (Table 6.3, Fig. 6.15) hosts at least five transiting exo-planets. Kepler-20 e is most likely a rocky planet made of iron and silicates. It has approximately the size of Venus and is the first discovery of a sub-Earth size planet by the KEPLER team. It is not habitable because the planet is on an orbit very close to its host star and the equilibrium temperature is as high as 1000 K. Two planets, one Earth-sized ($1.03 R_\oplus$) and the other smaller than the Earth ($0.87 R_\oplus$), orbiting the star Kepler-20, which is already known to host three other, larger, transiting planets were reported by [34]. The gravitational pull of the new planets on the parent star is too small to measure with current instrumentation. The authors applied a statistical method to show that the likelihood of the planetary interpretation of the transit signals is more than three orders of magnitude larger than that of the alternative hypothesis that the signals result from an eclipsing binary star. Theoretical considerations imply that these planets are rocky, with a composition of iron and silicate. The outer planet could have developed a thick water vapor atmosphere.

How can spectral lines emitted by a planetary atmosphere be distinguished from (a) stellar lines (b) line originating the Earth' atmosphere?

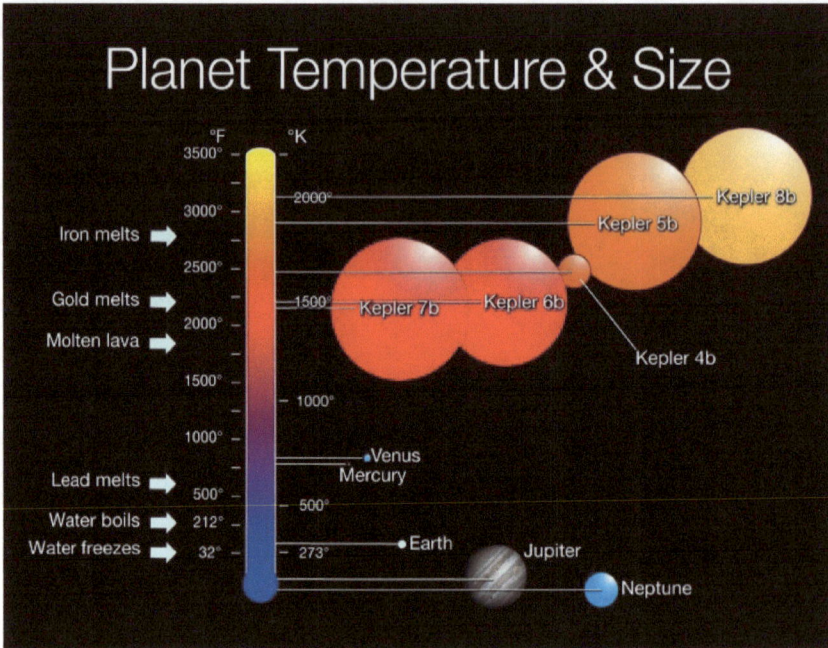

Figure 6.16: Examples of the first exoplanets detected by KEPLER. Credit: JPL

Table 6.3: The Kepler-20 planetary system.

	Kepler -20b	Kepler -20e	Kepler -20c	Kepler -20f	Kepler -20d
Semi Major Axis [AU]	0.04537	0.06335	0.093	0.1378	0.3453
Period [days]	3.69612	6.09849	10.8541	19.5771	77.6118
Mass [M_{Jup}]	0.027	0	0.051	0	0.06
Radius [R_{Jup}]	0.17	0.0791	0.27	0.093	0.25
Eccentricity	0	0	0	0	0
Inclination [deg]	86.5	87.5	88.39	88.68	89.57

Activities

Compare the habitability on moons orbiting a) a hot Jupiter, b) a hot Neptune. What is the amount of tidal heating to be expected on such moons?

Send Orders of Reprints at bspsaif@emirates.net.ae

CHAPTER 7

Host Stars of Planetary Systems

Abstract: In this chapter we will give an overview on the different types of host stars of planetary systems. The Hertzsprung Russel diagram showing the physical properties of stars and their relation is reviewed. Most stars can be found on the main sequence in this daigram. We concentrate on main sequence stars of spectral classes F, G, K, and M. Their main sequence lifetimes are larger than 1 Gyr. This time range was needed on Earth for life to develop and seems adequate for life to evolve on possible planetary systems around other stars[1].

The main message given in this chapter is that possible host stars of earthlike planetary systems might be the majority of all stars since the must be low mass stars like our Sun. However, we do not know yet the percentage of habitable earthlike planets around them.

Keywords: Exoplanets: host stars; Main Sequence; Main Sequence lifetime; Hertzsprung Russell diagram; Life: develop time

7.1   Main Sequence Stars

7.1.1   Stellar Parameters

Stars are characterized by their mass, temperature, age, chemical composition, magnetic fields, rotation and further parameters. The most important parameter that determines the stellar evolution is the stellar mass. Accurate stellar mass determination is only possible when a star has a companion. In such a case, the mass of a star M_1 follows from Kepler's third law. M_2 is the mass of a body (planet or small companion star) that orbits about the star, a being the semi major axis of that orbit and T the orbital period of M_2:

$$\frac{a^3}{T^2} = \frac{G}{4\pi^2}(M_1 + M_2) \qquad (7.1)$$

Stellar masses M_* are found in the range $0.08 \mathrm{M}_\odot < \mathrm{M}_* < 50 \mathrm{M}_\odot$, where $\mathrm{M}_\odot = 2 \times 10^{30}$ kg denotes the mass of the Sun. The mass of a star is related to its total energy production and therefore to its luminosity. For main sequence stars there exists a simple relation between mass and luminosity L:

$$L = M^{3.5} \qquad (7.2)$$

Stellar luminosity is related to its temperature T and surface, determined by the radius R:

$$L = 4\pi R^2 \sigma T^4 \tag{7.3}$$

Other important stellar parameters are:

- Stellar distances: considered as fundamental but not an intrinsic parameter. Stellar distances can be measured by determining their parallax, being the angle where Earth's orbit could be seen from a star. This defines the astrophysical distance in a unit called *parsec*. A star is said to be at a distance of 1 parsec if its parallax is $1''$. 1 pc = 3.26 Ly.

- Stellar radii: once the apparent diameter of a star is known its real diameter follows from its distance d. The problem is to measure apparent stellar diameters since they are extremely small. One method is to use interferometers, and the other one is to use occultation of stars by the Moon or mutual occultations of stars in eclipsing binary systems. All these methods have been described in general textbooks about astronomy.

- Once mass and radius are known, the density and the gravitational acceleration follow them. These parameters are important for the stellar structure.

- Stellar rotation: For it can be assumed that a star consists of two halves, one half approaches to the observer and the spectral lines from that region are blueshifted, while the other half moves away and the spectral lines from that area are redshifted. The line profile we observe in a spectrum is a superposition of all these blue- and redshifted profiles where rotation causes a broadening of spectral lines;

- Stellar magnetic fields: they will be discussed in more detail when considering the Sun. Magnetically sensitive spectral lines are split into several components under the presence of strong magnetic fields.

What would be the orbital period of a hot Jupiter at distance 0.01 AU orbiting a one solar mass star?

7.1.2   The Hertzsprung Russell Diagram, HRD

At the beginning of the twentieth century Hertzsprung and Russell came up with the idea to plot stellar luminosity against their temperature or spectral type (Fig. 7.1). They found that stars do not appear randomly distributed in such a diagram but more than 90% of all stars are found on a diagonal which they named as the main sequence. The temperature of a star depends on its color: blue stars are hotter than red stars. In the HRD the hottest stars are found to be present on the left side since the temperature increases from right to left.

Stellar brightness is given in *magnitudes*. The magnitude scale of stars was chosen such that a difference of 5 magnitudes corresponds to a factor of a 100 times in brightness. The smaller the number (which can also be negative) the brighter will be the star. The brightest planet Venus *e.g.* has a magnitude $-4.^{m}5$ and the Sun has $-26.^{m}5$. The faintest stars being visible to the naked eye have magnitude $+6.^{m}0$. Since the apparent magnitudes depend on the intrinsic luminosity and the distance of a star absolute magnitudes were invented:

the absolute magnitude of a star (designated by M) is the magnitude a star would be at a distance of 10 pc. In the HRD we can plot absolute magnitudes as ordinates instead of luminosities. The relation between m and M is given by:

$$m - M = 5 \log r - 5 \tag{7.4}$$

This equation tells us, that by knowing the intrinsic brightness of a star, or its absolute magnitude, we can find the distance of the star, r by comparing the absolute magnitude with the easily measurable apparent magnitude.

Calculate the absolute magnitude of the Sun

Stars can be considered to a very good approximation as *black body* radiators. A black body is a theoretical idealization: an object that absorbs completely all radiation at all wavelengths. The radiation of a black body at a given temperature is given by the *Planck law* (see also Fig. 7.2):

$$I_\nu = B_\nu = (2h\nu^3/c^2)/\exp(h\nu/kT_S) - 1 \tag{7.5}$$

Thus it depends only on the temperature T_S of the object. Here, I_ν is the intensity of radiation at frequency ν; h, k, c are Planck's constant, Boltzmann's constant and the speed of light respectively. $h = 6.62 \times 10^{-34} \,\mathrm{Js}^{-1}$, $k = 1.38 \times 10^{-23} \,\mathrm{JK}^{-1}$. If that equation is integrated over all frequencies (wavelengths), a formula can be obtained for the total power emitted by a black body, Stefan-Boltzmann law:

$$\int_0^\infty B_\lambda d\lambda = \sigma T^4, \tag{7.6}$$

and for the luminosity of a star with radius R:

$$L = 4\pi R^2 \sigma T_{\mathrm{eff}}^4 \tag{7.7}$$

For the Sun $T_{\mathrm{eff}} = 5\,785\,\mathrm{K}$. This formula defines the effective temperature of a star $\sigma = 5.67 \times 10^{-8} \,\mathrm{W/m^2 K^4}$ is the Stefan Boltzmann constant.

By taking the derivative with respect to λ of Planck's Law and setting it equal to zero, one can find the peak wavelength, where the intensity is at maximum:

$$T\lambda_{\max} = 2.9 \times 10^{-3} \,\mathrm{m\,K} \tag{7.8}$$

This is also called *Wien's law.*

From Wien's law we see that blue stars must be hotter than red stars (because $\lambda_{\max,\mathrm{blue}} < \lambda_{\max,\mathrm{red}}$).

What is the $\lambda_{\max}$ of radiation emitted from a cold planet at a temperature of 300 K? Compare this radiation with the radiation of the Sun at this wavelength!

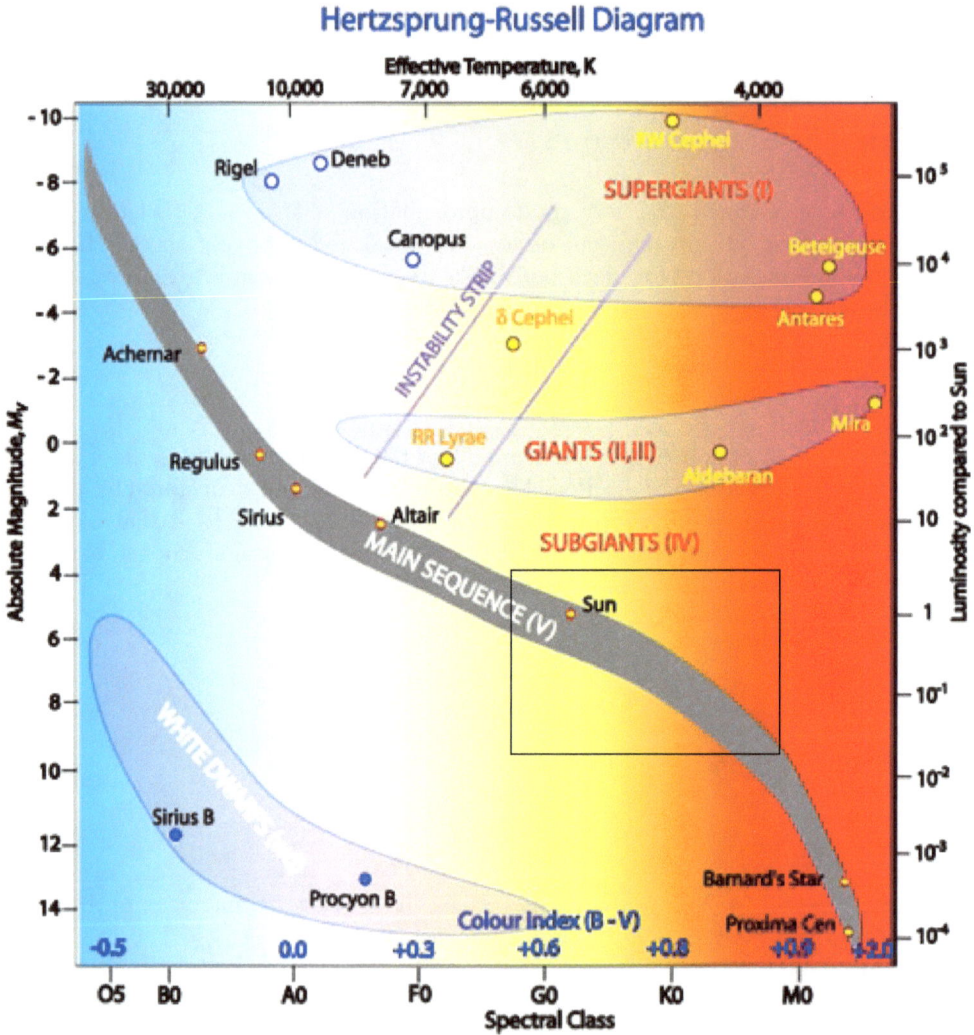

Figure 7.1: The Hertzsprung-Russell diagram. The main sequence stars that are of interest to astrobiology are marked by an rectangle.

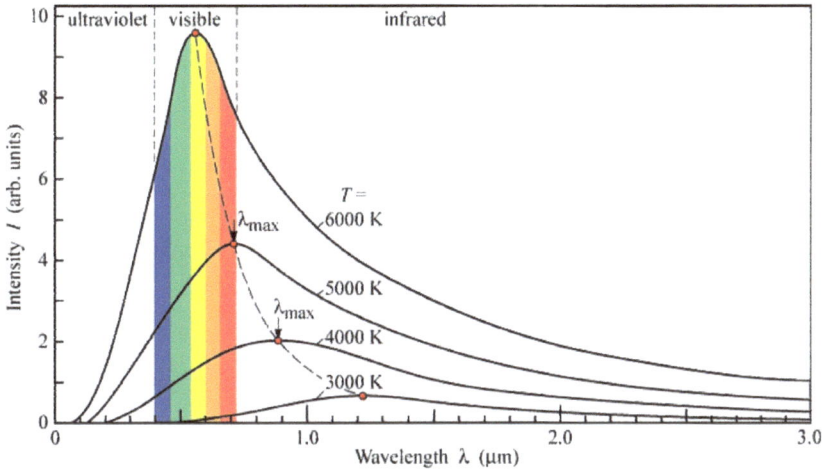

Figure 7.2: Planck curves for star at different temperatures. Note that the maximum amount of emitted radiation moves to shorter wavelengths for higher stellar temperatures.

7.1.3 Stellar Evolution

Stars are not randomly distributed in the HRD:

- Main sequence stars: most stars are found along a diagonal from the upper left (hot) to the lower right (cool).

- Giants, supergiants: they have the same temperature as that of the corresponding main sequence stars but are much brighter having larger diameters (see equation 7.7).

- White dwarfs are faint but very hot objects thus from their location at the lower left in the HRD it seems that they must be very compact (about 1/100 the size of the Sun).

Why most of the stars we observe lie on the Main sequence? The answer is quite easy: because this denotes the longest phase in stellar evolution. Let us discuss this briefly for the Sun:

The lifetime of a star depends on the amount of fuel × the mass of the star divided by the rate fuel is used × the luminosity of a star.

For the main sequence lifetime a formula can be given:

$$\tau_{\rm MS} = (1.0 \times 10^{10}) \times \frac{M}{(M_\odot)} \frac{L_\odot}{(L)} \sim 10^{10} \left(\frac{M}{M_\odot} \right)^{-2.5} \quad {\rm yr} \qquad (7.9)$$

The main steps in the evolution of the Sun are:

- Pre main sequence evolution: the Sun was formed from a protostellar gas and dust cloud and before it could reach the main sequence where it spends most of its life, the contracting Sun passed a violent youth, the T Tauri phase.

- At the main sequence the Sun changes extremely slowly remaining there about 10^{10} years. In the core H is transformed to He by nuclear fusion.

Table 7.1: Spectral classification of stars

O	ionized He, ionized metals
B	neutral He, H stronger
A	Balmer lines of H dominate
F	H becomes weaker, neutral and singly ionized metals
G	singly ionized Ca, H weaker, neutral metals
K	neutral metals molecular bands appear
M	TiO, neutral metals
R,N	CN, CH, neutral metals
S	Zirconium oxide, neutral metals

- The Sun evolves to a red giant, that will expand and the Earth will become part of the solar atmosphere. The expansion starts when all H is transformed to He in the core. Then a H burning shell supplies the energy. The He flash sets in as soon as He burning starts in the center. The Sun will evolve to a red giant for some 10^8 yrs. It will extend beyond the Earth's orbit.

- Finally, the Sun becomes a white dwarf which slowly cools.

During its evolution, the Sun dramatically changes its radius (the subscript $\odot$ denotes the present day value):

1 $R_\odot$ (present Sun) $\rightarrow \sim 10^4\,R_\odot$ (red giant), $\rightarrow 0.01 R_\odot$ (white dwarf)

For space weather long term evolutionary effects are negligible. But it is interesting to investigate them especially for the early Sun (see the chapter on the faint young Sun problem).

For the main sequence stars there exists a relation between their mass and luminosity:

$$L \sim M^{3.5} \tag{7.10}$$

From 7.10 we see that more massive stars are very luminous. Therefore, they use up their nuclear fuel much more rapidly than low mass stars like our Sun. Massive main sequence stars those being observed today must have been formed in very recent astronomical history[2].

Compare the luminosities for (a) a star with one solar mass, (b) a star with 10 solar masses.

7.1.4   Spectral Classes

According to their spectra, stars can be classified in the following sequence: O-B-A-F-G-K-M (see Fig. 7.3). This is a sequence of temperature (see Table 10.10): O stars are the hottest ones, M stars the coolest; the number of absorption lines increases from O to M. Some characteristics are described in Table 7.1.

In Table 7.2 the effective temperature of stars at the corresponding spectral type is listed.

The luminosity of a star depends on a) temperature $\sim T^4$, b) surface which is $\sim R^2$. Since e. g. a K star may be a dwarf main sequence star or a giant, luminosity classes

---

[2]In some large interstellar nebulae one observes stars that have an age of some 10^5 years

Table 7.2: Effective Temperature as a function of spectral type

Spectral Type	O	B0	A0	F0	G0	K0	M0	M5
T_{eff} [K]	50 000	25 000	11 000	7 600	6 000	5 100	3 600	3 000

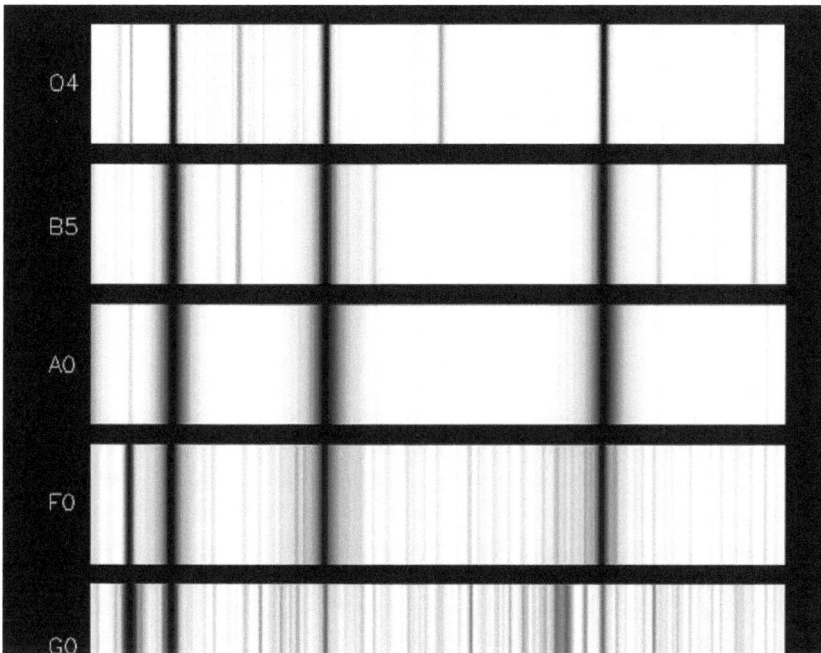

Figure 7.3: Spectral classification of stars. Note the increase of lines in the spectra of stars of type G, K and M. Credit :http://cde.nwc.edu

Table 7.3: Stellar parameters and their influence on habitable zones

Parameter	Influence
Age	Important for total lifetime of star
Mass	Determines the lifetime; stars with high masses evolve too fast
Temperature	Coupled to the mass
Composition	Nearly identical for all stars
Magnetic field	Important; determines stellar activity
Rotation	Important; determines stellar activity

have therefore been introduced for the classification. Class I contains the most luminous supergiants, whereas class II the less luminous supergiants. Class III are the normal giants, class IV the sub giants and class V the main sequence.

Consider a main sequence O5 star. It has about 60 times the mass of the Sun but is is about 800 000 times as luminous as the Sun. Therefore from the equation 7.9, we can calculate its main sequence lifetime:

$$\tau_{\mathrm{MS}}(\mathrm{O5}) = (1.0 \times 10^{10})\frac{60}{800,000} \sim 8 \times 10^5 \ \mathrm{yr} \tag{7.11}$$

Since their lifespan is less than 1 million years several generations of such stars have existed and died during the evolution of the universe (which has started 13.5 Billion yrs ago). On the other hand low mass stars live much longer. A K5 star with a mass of 0.67 $M_\odot$ and a luminosity of $0.15\,L_\odot$ has a lifetime of 5.3×10^{10} years, 5 times that of the Sun.

This phenomenon provides important consequences for our study. A star with twice the Sun's mass would survive only about a billion years on the main sequence where it remains stable. Since on Earth it probably took about that time for the first lifeforms, such a short stellar lifetime is hardly enough time for anything more than most of the other primitive lifeforms to evolve.

Calculate the main sequence lifetime for a star (a) with one solar mass and (b) with 1.5 solar masses.

7.1.5   Stellar Activity

When the host star provides the energy source for life, the habitability on planets around that star becomes strongly dependent on stellar physical parameters like: (i) stellar mass, (ii) stellar temperature, (iii) age of the star, (iv) stellar activity and others.

There are relations between these parameters. The activity of a star depends on its rotation and magnetic fields. The lifetime of a star depends strongly on its mass. A summary of stellar parameters and their importance for habitable zones around the stars is given in Table 7.3.

7.2   The Sun -The Star We Live With

Our Sun is the only star being close enough to observe details on its surface such as sunspots, faculae, prominences, coronal holes, flares *etc.*, all summarized as solar activity phenomena. Therefore, the study of the Sun is important for astrophysics in general. Theories about stellar structure and evolution can be studied in detail on the Sun [3].

7.2.1   Basic Properties

As it has been mentioned, the Sun is a G2V star in the disk of our Galaxy. The mass of the Sun is:

$$M_\odot = 1.99 \times 10^{30}\,\text{kg} \tag{7.12}$$

An application of Kepler's third law gives us the mass of the Sun if its distance is known which again can be derived from Kepler's third law:

$$\frac{a^3}{P^2} = \frac{G}{4\pi^2}(M_1 + M_2) \tag{7.13}$$

In the above case a denotes the distance Earth-Sun (150×10^6 km), P the revolution period of the Earth around the Sun (1 year), M_1 the mass of the Earth and M_2 the mass of the Sun. One can make the assumption that $M_1 << M_2$ and therefore $M_1 + M_2 \sim M_2$.

If we know the distance of the Sun and its angular diameter the solar radius is obtained:

$$R_\odot = 6.96 \times 10^8\,\text{m} \tag{7.14}$$

The measurement of the Sun's angular diameter is not trivial; one possibility is to define the angular distance between the inflection points of the intensity profiles at two opposite limbs. Such profiles can be obtained photoelectrically and the apparent semi diameter at mean solar distance is about 960 seconds of arc ($''$). The orbit of the Earth is elliptical and at present, perihelion (smallest distance of the Sun) is in January.

Knowing the mass and radius of the Sun, the mean density can be calculated:

$$\bar{\rho} = 1.4\,\text{g/cm}^3 \tag{7.15}$$

The gravitational acceleration is given by:

$$g = GM/R^2 = 274\,\text{m/s}^2 \tag{7.16}$$

The *solar constant* has been defined as the energy crossing unit area of the Earth's surface perpendicular to the direction from the Earth to the Sun in every second. In SI the units are $\text{W\,m}^{-2}$. UV and IR radiations from the Sun are strongly absorbed by the Earth's atmosphere. Therefore, accurate measurements of the solar constant have to be done with satellites. ACRIM on SMM and ERB on Nimbus 7 showed clearly that the presence of several large sunspots which are cooler than their surroundings depress the solar luminosity by $\sim 0.1\%$. The Variability IRradiance Gravity Oscillation (VIRGO) experiment on the SOHO satellite is being performed to observe total solar and spectral irradiances at 402 nm (blue channel), 500 nm (green channel), and 862 nm (red channel) since January 1996 (for a review see *e.g.* Pap *et al.* (1999) [77]). The solar luminosity is:

---

[3]see *e.g.* The Sun and Space Weather, A. Hanslmeier, 2007, Springer

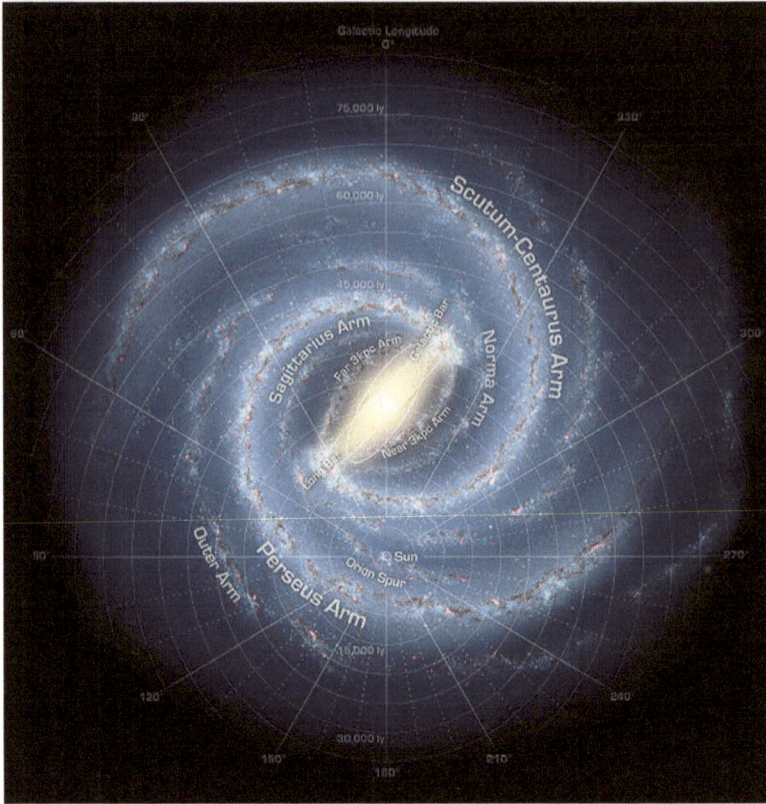

Figure 7.4: Location of the Sun in the Galaxy.

$$L_\odot = 3.83 \times 10^{26}\,\text{W} \qquad (7.17)$$

And the effective temperature:

$$T_{\text{eff}\odot} = 5780\,\text{K} \qquad (7.18)$$

7.2.2 Location of the Sun

Our Solar System is located in the Milky Way Galaxy. Our Galaxy contains more than 2×10^{11} solar masses (*i.e.* at least as many stars). The mass of the Galaxy can be inferred from the rotation of the system. All stars rotate about the center of the Galaxy which is at a distance of about 27 000 light years (Ly) to us [4].

At the location of the Sun in the Galaxy (Fig. 7.4), one period of revolution about the galactic center takes about 220 Million years. Galaxies in general contain some 10^{11} stars.

---

[4]1 Ly $= 10^{13}$ km, the distance light travels within one year propagating through space at a speed of 300 000 km/s

About 50% of the stars have one or more stellar companions. The diameter of our Galaxy is about 100 000 Ly. Galaxies are grouped into clusters- our Galaxy belongs to the so called local group of galaxies. The small and large Magellanic cloud are two small dwarf galaxies which are satellites of our system. The nearest large galaxy is the Andromeda galaxy being at a distance of more than 2.4 Million Ly.

Many galaxies appear as spiral galaxies. Young bright stars are found in the spiral arms, older stars in the center and in the halo of a galaxy.

7.2.3   Anatomy of a Star

It is quite remarkable that the basic structure of stars can be understood by some simple equations. The hydrostatic equilibrium states that the force acting on a small element inside a star is equal to the mass times the acceleration due to gravity:

$$F = dPdA = -g\rho(r)dA \tag{7.19}$$

dA is a surface element, P denotes pressure and $\rho(r)$ the density at a distance r from the stellar center. Usually the equation of hydrostatic equilibrium is presented as

$$\frac{dP}{dr} = -g\rho(r) \tag{7.20}$$

The next fundamental equation is that of mass continuity:

$$\frac{dM}{dr} = 4\pi r^2 \rho \tag{7.21}$$

The gradient of luminosity is given by

$$\frac{dL}{dr} = 4\pi r^2 \rho\epsilon \tag{7.22}$$

where ϵ denotes the energy generation which mainly occurs in the stellar cores.

The temperature gradient is

$$\frac{dT}{dr} = -\frac{3\kappa L\rho}{16\pi acr^2 T^3} \tag{7.23}$$

κ is the *opacity* and measures the resistance of the material to energy transport.

In the core of stars, energy is generated by the fusion of two lighter particles forming a heavier particle whose mass is smaller than the mass of its constituents, the mass defect being transformed into energy according to $E = \Delta Mc^2$.

Let us consider the fusion of H into He. The mass of 4 H is[5]:

$$4 \times 1.008145 \, \text{AMU} \tag{7.24}$$

The mass of the resulting He atom is

$$4.00387 \, \text{AMU} \tag{7.25}$$

Thus the mass difference ΔM is

$$0.02871 \, \text{AMU} \sim 4.768 \times 10^{-29} \, \text{g} \sim 4.288 \times 10^{-12} \, \text{J} \sim 26.72 \, \text{MeV} \tag{7.26}$$

---

[5] 1 AMU= 1/12 of the mass of the Carbon isotope $^{12}\text{C} = 1.66 \times 10^{-27} \, \text{kg} = 931\text{MeV}/c^2$

and 0,7% of the mass is converted to energy by Einstein's relation[6]

$$E = mc^2 \tag{7.27}$$

Compare the efficiency of nuclear fusion with electron–positron annihilation!

How can these models be tested experimentally?

7.2.4   Probing the Solar Interior

Models of the Earth's interior predict how density and temperature change at different depths. These variations affect the propagation of pressure waves by bending the parts of the waves. Therefore, one can test models of the Earth's interior by comparing measurements of seismic waves from earthquakes with model predictions of how seismic waves should travel.

Solar Oscillations

The Sun, as well as stars oscillates, vibrates or rings something like a struck bell. However, the vibrations of the Sun are very complex, with many different frequencies of vibrations occurring simultaneously. By analyzing the different oscillation modes one can infer the internal variation of temperature, pressure, density and even rotation. This technique is called *Asteroseismology* or *Helioseismology* when applied to our sun. The Global oscillation Network Group, GONG is a network of six solar observation stations spread around the globe and the Sun is observed almost continuously. Another project to measure these oscillation is the Solar and Heliospheric Observatory, an ESA-NASA mission located at the Lagrangian point L_1 approximately 1,500,000 km from Earth directly between Earth and Sun. Oscillations can occur in radial, in azimuthal directions or in directions vertical to the line of sight (these can be only measured for the Sun because it requires a spatially resolved object and stars always appear - even in the largest telescopes - as point like sources.

Solar Neutrinos

Another way to observe the stellar interior is to count the number of neutrinos emitted during the nuclear fusion reactions. Neutrinos are electrically neutral particles with a very small rest mass. They very weakly interact with matter only, so they can penetrate the whole stellar body without any interaction. Neutrinos are produced by the thermonuclear fusion reactions described above. The first experiment designed to detect solar neutrinos consisted of a tank filled with several 100,000 l of a cleaning fluid (perchlorethylene, C_2CL_4). This tank was placed within the Homestake gold mine buried 1500 m deep. Very few neutrinos interact with the chlorine atoms and

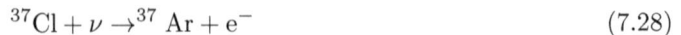

$$^{37}\mathrm{Cl} + \nu \rightarrow ^{37}\mathrm{Ar} + \mathrm{e}^- \tag{7.28}$$

To illustrate how difficult it is to detect solar neutrinos consider that within two days 10^{22} neutrinos pass through the detector but only one neutrino interacts with the chlorine atom!

The Super Kamiokande experiment is located in an active Zinc mine 2700 m under Mount Ikena near Kamioka, Japan. 50,000 tons of pure water are surrounded by 13,000

---

[6]1 eV=1.6 × 10^{-19} J

photomultiplier tubes capable of registering faint flashes of light that are produced by interaction of neutrinos with water. Other experiments use Gallium (SAGE experiment and GALLEX experiment) where the reaction

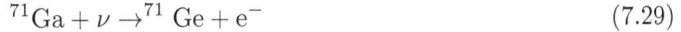

$$^{71}\text{Ga} + \nu \rightarrow ^{71}\text{Ge} + e^- \tag{7.29}$$

is measured.

The amount of solar neutrinos passing through the human body has been calculated 400 trillion per second. Measuring solar neutrinos proves that at the solar core nuclear reaction really works but after the initial joy at this confirmation, it became evident that only one third of the expected neutrinos were measured. This was called the solar neutrino problem. Was the understanding of the structure of the Sun wrong? It turned out that the discrepancy between experiment and neutrino fluxes predicted from models can be explained by neutrino oscillations. If neutrinos have a tiny amount of mass[7] they can oscillate between three different flavors: there exist electron neutrinos, muon neutrinos and tau neutrinos. The first neutrino detectors were able only to detect electron neutrinos.

This is a remarkable example how astrophysical observations and measurements can lead to new concepts and ideas in physics in general.

7.2.5   The Solar Interior

The solar interior consists of several layers (see also Fig. 7.5):

- core: in the core the thermonuclear fusion processes occur. It extends from 0 to 0.3 $R_\odot$, $R_\odot = 6.96 \times 10^5$ km, being the solar radius.

- radiative zone: the energy is transported by radiation, *i.e.* by numerous absorption - reemission processes of photons. This zone extends from 0.3 to about 0.7 $R_\odot$.

- convective zone: here the energy is transported to the surface by convective motions. Hot gas moves upwards, cools down and sinks back again. It extends from about 0.7 to 1.0 $R_\odot$.

- tachocline: between the radiative zone and the convective zone there is a small layer called the tachocline. It is assumed to be the layer where the solar dynamo has its origin.

At the surface, the Sun rotates differentially, near the solar equator one rotation takes about 25 days, while near the poles about 34 days. This differential rotation pattern persists down to the tachocline. In layers deeper than the tachocline the Sun rotates like a rigid body. This sharp transition between differential rotation and solid rotation causes shear stresses and generates currents and magnetic fields.

7.2.6   The Solar Atmosphere

The solar atmosphere extends from what is called the solar surface (where the optical depth reaches unity and no radiation escapes) to the solar corona extending over several solar radii and even the heliosphere which engulfs the whole planetary system.

---

[7]The neutrino was long thought to have zero mass and to travel with the speed of light

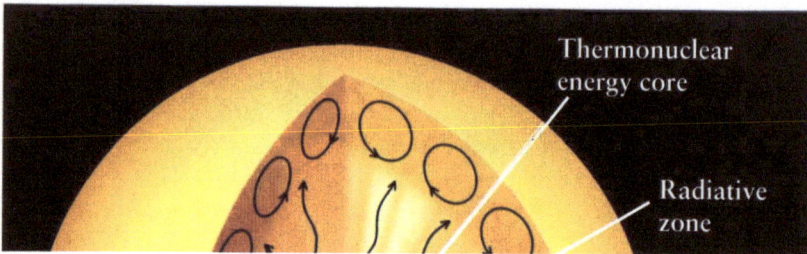

Figure 7.5: Interior structure of the Sun. Credit: http://crab0.astr.nthu.edu.tw/

- Photosphere: more than 99 % of visible solar radiation comes from the photosphere. It is a very thin layer of about 400 km therefore the solar limb appears relatively sharp. Well known phenomena in the photosphere are granulation and sunspots. The granulation results from overshooting convective motions and is a cellular like structure, the cells have a diameter of about 1000 km. Sunspots are regions of intense magnetic fields inhibiting convective energy transport to the surface. In the photosphere the temperature is about 5800 K, while in the dark umbrae of sunspots it is about 2000 K lower.

- Chromosphere: the temperature in the photosphere reaches a minimum at a height of about 400 km and then sharply rises in the chromosphere which can be observed during a total solar eclipse or in certain wavelengths, such as the prominent Hα line where the temperature reaches almost 10,000 K.

- Corona: in the transition region between the chromosphere and corona the temperature rises again very sharply reaching up to 1 Million K in the corona. Here the magnetic fields dominate the plasma motions and loop like structures of plasma that follow magnetic field lines can be observed. Magnetic fields can reconnect reaching at a less energetic state where energy is released in form of radiation and particles. When these phenomena occur in the lower corona/upper chromosphere we call solar flares, when they occur in the higher corona we mean coronal mass ejections. Typical energies released during a flare within about half an hour are in the range of several million hydrogen bombs.

- Solar wind: this is a continuous stream of charged particles. The fast component (velocities up to 800 km/s) emitted in so called coronal holes can be seen as dark areas in X-ray images of the Sun. X rays are emitted from the corona. In the coronal holes since the temperature is lower, they appear dark in X-rays and the particles can escape from open magnetic fields.

- Heliosphere: solar wind and solar magnetic fields define the heliosphere that engulfs the whole planetary system. The expansion of the heliosphere is about 100 AU. The heliopause is defined as the region where the solar wind pressure reaches the pressure of the interstellar medium through which the Sun is travelling. At the beginning of the 21st century the Voyager 1 spacecraft has succeeded in reaching this outer edge of the Solar System. Since charged incoming cosmic ray particles are deflected by the heliopause, their number is strongly reduced when the solar wind is strong *i.e.* during high periods of solar activity.

7.2.7   The Active Sun

Telescopic observations of sunspots (Fig. 7.6) date back for almost 400 years. There exists an 11-year sunspot cycle, the cycle neither being perfectly periodic nor constant is the amplitude of the maxima. The time between peaks in the number of sunspots varies between 9.7 and 11.8 years. There were also times when sunspot activity completely disappeared (*e.g.* during the Maunder minimum which lasted from 1650 to 1715). In the early 20th century, G. Hale discovered that in one sunspot cycle the leading sunspot in each pair tends to be a north magnetic pole. In the next sunspot cycle this polarity is reversed. Therefore, the magnetic solar activity cycle continues for 22 years. Sunspots are darker than the surrounding photosphere so one might assume that during the phase where large spots are visible on the Sun, less solar radiation is emitted towards Earth. But sunspot groups are very

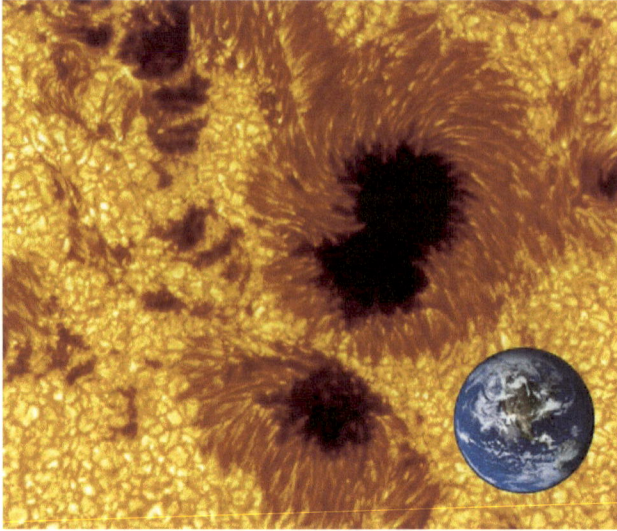

Figure 7.6: Sunspot group and for comparison of the Earth. Note also the cellular like pattern which is called granulation is a result of the convective energy transport to the solar surface. Courtesy: Randy Russell using images from the Royal Swedish Academy of Sciences (sunspot image) and NASA (Earth image)

often accompanied by brightenings in the chromosphere which can be seen in Hα. So the sunspot deficit is overcompensated by these brightenings (faculae, chromospheric network) and the Sun radiates more energy during sunspot maxima.

Does the Sun emit more radiation when it is heavily spotted?

Imagine 0.1% of the solar surface is covered by sunspot groups where the temperature is lower by about 1500 K with respect to the surroundings. What does it mean for the total solar luminosity?

Prominences are magnetic flux tubes of relatively cool gas (5,000 to 10,000 K) extending through the million K corona. The loop like structures are anchored in the photospheric active regions. Prominences are often quiescent but can also erupt throughout the corona ejecting material at velocities of 1,000 km/s.

Not only the number of sunspots varies with that period but all other solar activity phenomena like the number of flares and CMEs (Figs. 7.7, 7.8). Flares emit short wavelength radiation which influences on the upper Earth's atmosphere. The ionization rate increases, ozone production is influenced and the radiation may be hazardous to astronauts in space. All these effects are called space weather.

Solar Satellite Missions

SOHO, the Solar and Heliospheric Observatory, is a project of international collaboration between ESA and NASA to study the Sun from its deep core to the outer corona and the

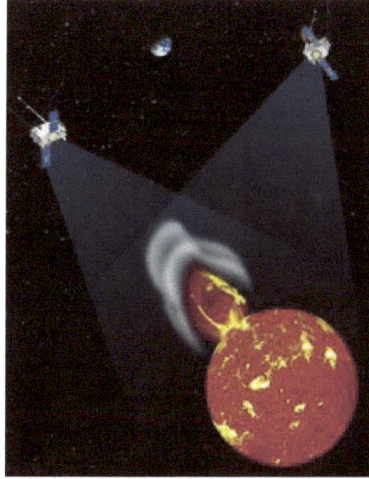

Figure 7.7: Observation of the Sun from Space with two satellites. This enables a stereo-scopic view of coronal massejections and other active phenomena on the Sun. Credit: NASA, STEREO.

solar wind. SOHO was launched on December 2, 1995. The Extreme ultraviolet Imaging Telescope (EIT) is an instrument on the SOHO spacecraft used to obtain high-resolution images of the solar corona in the ultraviolet range. The EIT instrument is sensitive to light of four different wavelengths: 17.1, 19.5, 28.4, and 30.4 nm, corresponding to light produced by highly ionized iron (XI)/(X), (XII), (XV), and helium (II), respectively. An example of a CME observed at 30.4 nm is given in Fig. 7.8. Other satellite missions to observe the Sun are STEREO (Solar TErrestrial RElations Observatory, launched Oct. 2006, Fig. 7.7), SDO (Solar Dynamics Observatory, launched Feb. 2010. Solar storms cause aurorae and can even disrupt power grids on Earth. Variations of the Earth's climate might be related to solar activity. Models show that the Sun could account for only 0.1 K differences in Earth's average temperature. This might not seem much but remember that triggering the onset of an ice age may require a drop in global temperature of only around 0.2 to 0.5 K.

The Early Sun

At present, the Sun might not be so active as to be a real threat to life on Earth. The young Sun, however, was much more active:

- X-ray and UV radiation: up to a factor of 1000 higher than today,

- activity: larger amplitudes, less periodic,

- the young Sun was no star to live with, its strong short-wavelength radiation was lethal to any lifeforms.

The higher level of solar activity during the early solar evolution (up to an age of several 100 million years) was extremely important for the formation of the early planetary atmospheres. The high level activity can be explained by a rapidly rotating early Sun. There is a well known relation in astrophysics between:

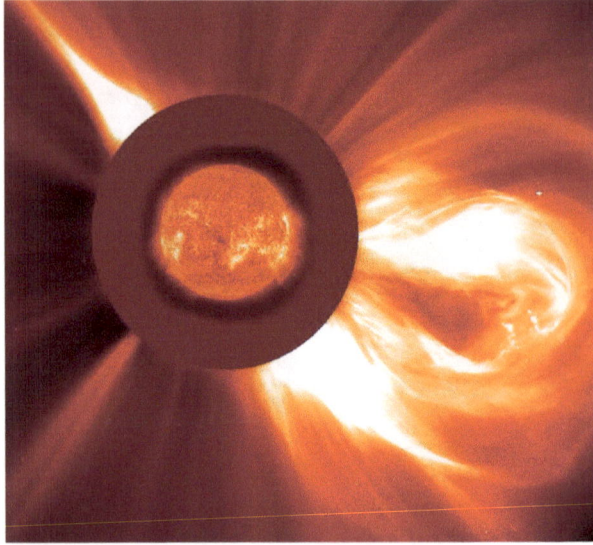

Figure 7.8: Observation of the Sun from Space in with the EIT instrument onboard SOHO satellite in the UV. A ring around the Sun is occulted, on the solar disk itself fine structures are seen. A large coronal mass ejection is seen on the right side. Credit: NASA, GSFC.

$$stellar\ rotation\ rate \rightarrow stellar\ activity \rightarrow stellar\ age.$$

As stars become older their rotation slows down because of removal of angular momentum due to solar wind loss.

7.3 Solar Like and Late Type Stars

As it has been discussed, the evolution of life even to very primitive forms lasted only about 1 Gyr on Earth. Therefore, the search for habitable planets around stars has to be limited to the stars with main sequence lifetime > 1 Gyr. Therefore, the following spectral classes of stars are of interest: F, G, K and M. Habitable zones around these stars will be discussed later. Here, we give a brief overview of the different spectral classes of stars that may act as host stars of habitable planets.

7.3.1 Late Type Stars

By the term late type star usually stars later than spectral type G are denoted in astrophysics. Here we define late type stars as stars of spectral types K and M. On the Hertzsprung-Russell diagram they appear at the right hand side of the Sun. As it has been argued already, for studies about habitability it is reasonable to concentrate on main sequence type stars, we will only briefly mention the giant and supergiant type stars. In Table 7.4 we give the spectral class, temperature, mass and approximated main sequence lifetime

Table 7.4: Solar-like and late type stars.

Spectral Type	Mass	Temperature	τ_{MS}
G	1.0	5,800	10×10^9
K	0.6	4,000	32×10^9
M	0.22	2,800	210×10^9

(MS lifetime, τ_{MS}) for these and solar like stars (spectral type G). Generally late type stars are cool stars and their masses are less than the mass of the Sun.

G stars are solar like and they appear as yellow stars whereas K stars appear orange. The spectra are dominated by H and K lines of potassium as well as lines of neutral Fe and Ti. Also molecular lines, such as CN and TiO are found because of the lower temperature than in the hotter stars.

The K type main sequence stars have masses between 0.5 and $0.8\,M_\odot$ and the temperature is between 3,900 to 5,200 K. Their luminosity ranges from 0.1 and $0.4\,L_\odot$.

7.3.2   M-Type Stars

Being the most important part of stellar population, the chances to detect an exoplanet around an M-type star are the highest and therefore they are treated separately here.

M type stars are cool red stars with a surface temperature less than 3,600 K. Molecular lines dominate their spectra. The masses are $< 0.5\,M_\odot$ and larger than the lower limit for stellar masses (about $0.08\,M_\odot$). The red M dwarfs have a low luminosity because of their relatively small size and temperature, $L < 0.08\,L_\odot$. M type giant stars have a large extension and luminosities $> 300\,L_\odot$ and M type supergiants have luminosities up to $10^5\,L_\odot$.

One famous example of an M type red dwarf is Proxima Centauri being the nearest star to the Solar System at a distance of 4.24 Ly. It is only visible in telescopes (apparent magnitude 11.1).

Barnard's star is of type M4Ve, the suffix e indicating emission lines. It is only about 18% the mass of the sun, and about 0.04% of the sun's luminosity and its age being 10 billion years. The current distance to the Sun is about 6 Ly but as it is approaching the closest distance will be 3.8 Ly in about 10 000 yr.

It seems that M-dwarfs constitute the most abundant stellar population in the universe: twenty of the 30 nearest stars to our planetary system are red dwarf stars. Since they constitute the oldest objects in the universe one would expect a very low metal content. However it has not been verified so far. Therefore, apart from their main sequence lifetime that is larger than the age of the universe they may also be stars of second generation. Shortly after the Big Bang massive stars may have formed, exploded right after several 10^5 years as supernovae and thus enriched the original composition of the universe which was only H and He by elements heavier than He.

One constraint for habitability around M-type stars is their activity. Many M stars exhibit high levels of activity (high XUV fluxes, powerful flares *etc.*) during extended periods of time that could be harmful to the evolution to life. A radial velocity accuracy of 3 m/s will permit the detection of super-Earths inside the habitable zone of stars later than M3, observing in the far IR is another possibility for their detection ([74]). M-Type Stars as Hosts for Habitable Planets were discussed in [76]. Habitable zones around M stars and the problem of small planetary magnetic fields because of tidal locking will be discussed later. M stars might emit strong coronal mass ejections and stellar winds depending on

Table 7.5: Early type stars.

Spectral Type	Mass	Temperature	τ_{MS}
O	32	35,000	10×10^6
B	6	14,000	270×10^6
A	2	8,100	800×10^6
F	1.25	7,000	4.2×10^9

their age. The convective zone reach deep into their interior. The chromospheric/coronal heating mechanism for M-type stars is different from that of early type stars. Being hotter, turbulent velocities on early type stars are higher than on late type stars. Higher turbulence favors models of magnetoacoustic heating whereas for late type stars this process becomes less important (some unknown X-ray heating mechanism of the corona is suggested) ([35]).

Planetary systems around late type stars and the habitable zone around these stars were also investigated ([55]).

7.3.3 Early Type Stars

All stars from spectral types O to F0-F5 are classified as early type stars. This is not related with their evolutionary status.

These are found on the left hand side of the HRD. Early type stars are more massive and luminous than the Sun, with a shorter main sequence lifetime. From Table 7.5 we see that only F-type and late A-type stars have main sequence lifetimes larger than 1 Gyr.

For our topic, the search for life in the universe we conclude that:

- early type stars evolve too fast so that higher life forms cannot evolve,

- early type stars emit a high portion in the UV range which is also destructive for life,

- early type stars play an important role for life since, because of their fast evolution, they enriched the universe with elements heavier than He. These elements (such as C, O, N,...) are essential for life.

What would be the space weather in the surroundings of early and late type stars?

7.3.4 Solar Twins

Solar twins stars are, basically the stars with the same mass, temperature, surface gravity, luminosity, metal content and age as the Sun. The main interest for these stars comes from the fact that planetary systems around them might be the best candidates for the search for extraterrestrial life. The existence of solar twins helps also to answer the question wether the Sun is a unique star.

A first systematic survey to reveal all solar twin stars within 50 pc of the Sun was made using the HIPPARCOS[8] astrometric and photometric database. 52 near main-sequence, G-type candidate stars with absolute magnitudes and color indices very closely approaching the Sun's were identified ([81]).

---

[8]High precision parallax collecting satellite, the mission operated between 1989 and 1993

18 Sco

One good candidate for a solar twin is 18 Sco. The star is at a distance of 45 Ly. In astrophysics all elements heavier than He are called metals. The metallicity of 18 Sco is 1.1 times that of the Sun, which means the abundance of elements other than hydrogen or helium is 10% greater. Its Lithium abundance is three times as high. Ca K observations of 18 Sco showed a pronounced rise suggesting that 18 Sco has a well-defined activity cycle that reached an apparent minimum in 1998 and showed a rapid rise through the most recent year 2000 data. The activity cycle of 18 Sco may be of greater amplitude than the Sun's and its overall chromospheric activity level is noticeably greater than the Sun's ([38]). The estimated cycle period is about seven years. Up to now the search for planets around 18 Sco was negative, radial velocity measurements have not yet revealed the presence of planets orbiting it.

Activities

Let us assume that the Sun has a mass of 10% larger. What would be the effects on its main sequence lifetime and luminosity. What would be the consequences for

- habitability on Earth
- habitability on Mars

Send Orders of Reprints at bspsaif@emirates.net.ae

CHAPTER 8

Habitable Zones about Stars and Planets

Abstract: To our knowledge life can only evolve in an environment that is quite similar to that on the Earth. Based on this unproven assumption, the habitable zone (HZ) has therefore been introduced assuming that life generally depends on such conditions, namely temperature, pressure, atmospheric chemistry similar to that on Earth.

A **habitable zone HZ** is defined as a region of space around a star or even around a planet, where the conditions are favorable for life based on

- complex carbon compounds,
- availability of fluid water.

Several authors call a planet or satellite of a planet habitable when liquid water can be found there. The liquid water belt concept was introduced by H. Shapley in the 1950s. By the 1970s M. Hart estimated extraterrestrial life as being extremely rare. This lead to the rare Earth hypothesis.

In the following sections we will outline that habitable zones do not only depend on the distance of a planet from its host star but also on the location of the planetary system in a galaxy as well as there might be habitable moons around giant planets.

The main message of this chapter is that there exist several habitable zones, circumstellar, galactic and even habitable zones around giant planets. Habitability however, strongly depends on the definition of life.

Keywords: Habitable zone; circumstellar habitable zone; Galaxy; galactic habiatble zone; Evolution of habitable zone; continuous habitable zone

Activities

Discuss why there is good reason to assume that life has evolved based on the same fundamental requirements found on Earth:

- Carbon based
- Liquid water

8.1   Circumstellar Habitable Zones

8.1.1   Extension of HZ

A habitable zone around a star must depend on stellar parameters such as stellar mass, luminosity, age, evolution as well as on parameters of the object where life is assumed to

Arnold Hanslmeier

exist.

What is the extension of an HZ around a star, and how it depends on the mass of a star and its age?

Let us consider first the radiative flux from a star similar to the Sun. The flux (integrated over all wavelengths) varies with the square of the distance from the star, and therefore:

$$F \sim 1/r^2 \tag{8.1}$$

Let us consider the case of the Solar System. To illustrate the different fluxes on our neighboring planets Venus and Mars, the following calculation has to be done: The semi major axis of Venus is 0.7 AU, the semi major axis of Mars is 1.4 AU. Then the solar flux at the orbit of Venus is:

$$F_{\text{Venus}} \sim 1/(0.7)^2 \sim 2 \tag{8.2}$$

On Venus, the solar flux to be expected will be twice than the flux on Earth; for Mars we obtain:

$$F_{\text{Mars}} \sim 1/(1.4)^2 \sim 0.5 \tag{8.3}$$

Therefore, flux changes by a factor of 4 between the orbits of Venus and Mars and the HZ even in case of a solar like star will be restricted to a small circumstellar region of about 0.7 AU extension (which is the difference between the semi major axis of Mars and Venus).

The luminosity of the Sun changed from 0.75 $L_\odot$ to its present value. At an age of only several hundred million years, the early Sun was fainter than the present Sun, so why there was water in liquid form even on early Earth, as we know it from sedimentary deposits? This is also known in literature as the faint young Sun problem. An increasing solar luminosity means that habitable zones (HZ) tend to move outward with time because the Sun, like all main sequence stars, has become brighter during its evolution. In order for life to evolve to higher forms, the planet must be continuously in the habitable zone that slowly progresses outward due to stellar evolution. This leads to the definition of a continuously habitable zone CHZ. The *continuously habitable zone* is the region in space where a planet remains habitable for some long time period τ_{hab}.

The evolution of intelligent life on Earth took $\sim 4\,\text{Gy}$ and this must be about τ_{hab}. Some other authors take smaller values such as $\tau_{\text{hab}}=3$ Gy or $\tau_{\text{hab}}=1$ Gy, for the evolution of microbiological life. From these values it becomes clear, that a planet must remain in a continuous habitable zone a considerable fraction of its lifetime in order for life to originate on his surface.

According to Hart, 1970, the inner and outer boundaries of the HZ are:

$$r_o/r_i \sim [L(3.5)/L(1.0)]^{1/2} \tag{8.4}$$

Here r_i denotes the inner boundary and r_o the outer boundary of a continuously habitable zone. $L(t)$ represents the luminosity after t billion years of a main sequence star of mass M.

For stars with different masses we see that:

- r_o/r_i becomes smaller for less massive stars

- It converges to $r_o = r_i$ for stars with $M \sim 0.83\,\text{M}_\odot$.

  This would implicate that low mass cool stars (of spectral type later than K1) have no continuously habitable zone. The probability to find a planet in such a narrow zone becomes quite low. This is now seen as less restrictive and HZ around late type stars is believed to exist. Since less massive stars provide the majority of stars in the universe, the chance to find life elsewhere becomes higher.

Up to now we have discussed radiation fluxes integrated over all wavelengths. For life, however, the more damaging part of radiation comes from the short wavelengths like UV and X-rays.

Circumstellar habitable zones (CHZ) strongly depend on UV radiation.

- On Earth, UV radiation between 200 and 300 nm is very damaging to biological systems.

- On the other hand it must be taken into account that UV radiation on the primitive Earth was one of the most important energy sources for the synthesis of biochemical compounds and therefore essential to biogenesis.

8.1.2   HZ and Planetary Atmospheres

Stars emit shortwavelength UV and X radiation. With no atmosphere, Earth would be a planet hostile to life. Planetary atmospheres have a strong effect on habitability. Consider the two extreme cases of a planetary atmosphere. Venus has an extremely dense atmosphere and the main constituent, CO_2 causes a strong greenhouse effect making that planet inhabitable. Mars has presently a low density atmosphere, therefore it is extremely cold.

The formation of CO_2 clouds plays a crucial role. Two boundaries can be defined:

1. Outer edge: this is defined by the distance of the planet from its central star at which the formation of CO_2 clouds starts. These clouds cool the planet's surface by increasing its albedo. Let $< CO_2 >$ denote atmospheric CO_2 concentration and T_{surf} denote the surface temperature on a planet, then:

$$< CO_2 > \sim 1/T_{surf} \tag{8.5}$$

2. Inner edge: because of the low distance of the planet to its central star, loss of water via photolysis occurs. Water goes into the stratosphere, the H_2O molecules are split by UV radiation from the star into H_2 and O and hydrogen, H_2, escapes from the atmosphere. This is also referred to as moist greenhouse effect.

$\rightarrow$If these constraints are applied to our Solar System, then the limits for the habitable zone are 0.95 AU$<$ HZ $<$ 1.37 AU.

It was also shown that the width of the HZ depends on the size of the planet. It increases for planets that are larger than Earth and it increases also for planets which have higher N_2 partial pressures.

In Fig. 8.1 the habitable zone is shown. for the Gliese system.

8.1.3   Circumplanetary HZ

So far HZs have been discussed arount a central star. If an HZ is defined by the presence of liquid water we must also include the interiors of giant planets and the ice covered Galilean Moons of Jupiter, the Martian subsurface and maybe even other places.

Therefore, the search for life must not be restricted to the study of circumstellar habitable zones. But it seems that life requires much more than just liquid water, *i.e.* an energy source.

In the cases of satellites of planets, such an energy source can be provided by other mechanisms than radiation from a central star. One possible mechanism in connection was already discussed: the tidal heating.

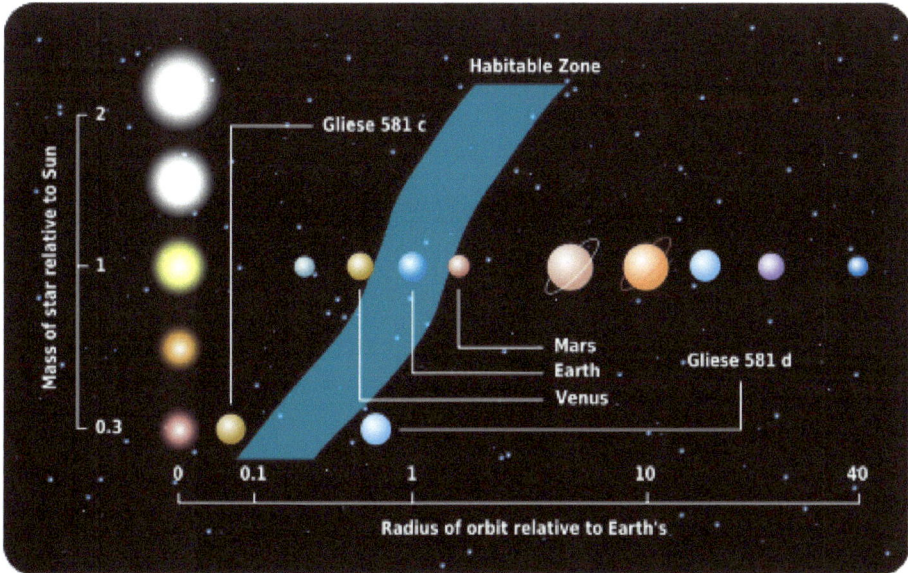

Figure 8.1: Comparison of the circumstellar habitable zone in our Solar System and the Gliese system. On the ordinate stars of different masses are shown. The habitable zone is shifted to larger distances from the central star for larger stellar masses. Credit: http://its-interesting.com

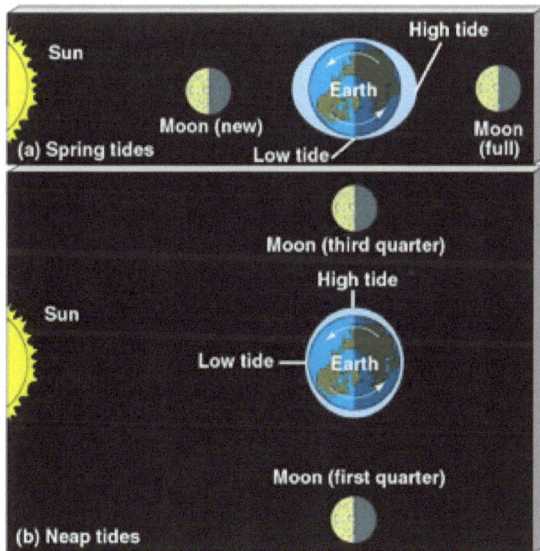

Figure 8.2: Tidal interaction between Earth and Moon.

8.2   Tidal Effects and Resonances

8.2.1   Tidal Locking

The rotation of a planet is an important factor in distributing heat from the sunlit side to the night hemisphere. Rotation and revolution of a planet around a star can be coupled, leading to a strong decrease of rotation rate and there changing the uniform distribution of heat on a planetary surface.

A well known example of tidal locking is the Earth-Moon system. From Earth, one can always see the same hemisphere of the Moon. It takes just as long to rotate around its own axis as it does to complete one revolution around the Earth also called *synchronous rotation*, the Moon is 1:1 tidally locked.

Let us consider a smaller body B which gets locked to a larger body A. The gravity of A produces a tidal force on B. This results in a deviation from its equilibrium configuration, *i.e.* B becomes slightly elongated and stretched along the axis oriented toward A and on the other hand it is compressed in the two perpendicular directions $\rightarrow$ tidal bulges (see Figure 8.2). The result is that B becomes an ellipsoid.

Two bodies A and B attract each other, $M_A > M_B$. The gravitational attraction of A on B is larger on the surface oriented towards A (*i.e.* $F_1 > F_2$)

Let us explain the two tidal bulges on Earth caused by the attraction of the Moon. F_2 denotes the force exerted by the moon on the face of the Earth towards it (*i.e.* the moon is in the highest point over the horizon there), F_z is the force that the moon exerts on the center of the Earth and F_1 is the force of the moon on the side opposite to it. Subtracting F_z from F_2 and F_1 we see that there is a resulting force that points away from the Earth's surface thus causing the tidal bulges.

This leads to the effect of bulge dragging:

- The material of B exerts resistance to this reshaping caused by the tidal force.

- There is also some delay: if the rotation period of B is shorter than its orbital period, the bulges are carried forward of the axis oriented towards A in the direction of rotation.

- The bulges lag behind it if the orbital period of B is shorter.

The bulges are displaced from the $A - B$ axis resulting in a torque. Note that there are in fact two torques which react in opposite sense since being not exactly equal:

- Torque on the A-facing bulge acts to bring B's rotation in line with its orbital period

- Torque on the side on B opposite to A acts in opposite sense

- The gravitational attraction on the A facing side is slightly higher than on the opposite side

- $\rightarrow$ tidal locking

Conservation of angular momentum in the system results into an additional effect.

- Because B slows down it loses angular momentum

- in order to conserve the total momentum, the momentum resulting from the orbital motion must increase

- $\to\to$ the semi major axis of the orbit of B increases

$$L_{\text{orbit}} = a_{\text{orbit}} M_B v_{\text{orbit}} \qquad (8.6)$$

Therefore the scenario for object B is as follows:

the rotational slowdown $\to$ increase orbital diameter.

when B starts off rotating too slowly, tidal locking speeds up rotation and lowers the orbit.

To Summarize, tidal locking leads to two effects depending on the initial rotation of B:

- $P_{\text{rot,B}} < P_{\text{orbit,B}}$: rotation slowdown, orbit increases,

- $P_{\text{rot,B}} > P_{\text{orbit,B}}$: rotation speedup, orbit decreases.

So far we have only considered the effects on the smaller body B. But the tidal locking effect of course also influences on the larger body A. The effects on A depend on the mass ratio M_B/M_A. For the *Earth-Moon* system this ratio is $1/81$, forcing the Earth to spin down gradually.

8.2.2   Orbit-Rotation Resonance

There also exists an orbit-rotation resonance. *Mercury* is in a 3:2 rotation-orbit resonance with the Sun which means that the planet makes 3 rotations after completing 2 orbits around the Sun.

The final configuration of such a tidally locked system is the one that occurs in the lowest energy. This means that the heavy side will face the planet when considering the case of a moon and a planet. The hemisphere that is facing the Earth is quite different – there are the large maria which are impacts of basins filled with lava later. The maria are composed of basalt and are heavier than the surrounding highland crust.

8.2.3   Tidal Locking Timescale

Let us consider the timescale for tidal locking:

$$t_{\text{lock}} \approx \frac{w a^6 I Q}{3 G M_p^2 k_2 R^5} \qquad (8.7)$$

w is the initial spin rate (revolutions per s), a the semi major axis of the satellite, $I \sim 0.4 M_S R^2$ the moment of inertia of the satellite with mass M_S, Q the dissipation function of the satellite, G the gravitational constant, M_P the mass of the planet, k_2 the tidal Love number, and R the radius of the satellite. The tidal Love number is:

$$k_2 \sim \frac{1.5}{1 + \frac{19\mu}{2\rho g R}} \qquad (8.8)$$

where, ρ is the density of the satellite, $g = GM_S/R^2$ the surface gravity of the satellite and μ the rigidity of the satellite. This value depends on the material which forms the satellite and the two extremes are:

- $\mu = 3 \times 10^{10}\,\text{Nm}^{-2} \to$ rocky satellites

- $\mu = 4 \times 10^9 \, \text{Nm}^{-2} \to$ icy satellites.

For the system Earth-Moon the ratio $k_2/Q = 0.0011$ can be taken. As a first approximation one can take $Q \sim 100$ and calculate k_2 according to the above given formula.

The formulae can be simplified with $k_2 << 1, Q = 100$ and assuming a revolution every 12 hours in the initial non-locked state. These values hold for asteroids in the Solar System because they have orbital periods between 2 h and 2 d. Then we arrive at:

$$t_{\text{lock}} \approx 6 \frac{a^6 R \mu}{M_s M_p^2} \times 10^{10} \quad \text{years}, \tag{8.9}$$

Masses are in kg, distances in m and μ in Nm^{-2}. Note that the tidal locking time strongly depends on the semi major axis of the satellite. Let us consider the factor R/M_s in the formula which is $R/(4\pi/3 R^3 \rho_s) \sim 1/R^2$. That means that larger satellites will lock faster than smaller ones[1].

8.2.4 Tidal Heating

Tidal forces result in deforming an object orbiting a larger one. Such a continuous compression produces heat.

Tidal heating occurs at a rate of:

$$F_{\text{tid}} = (9/19)\rho^2 n^5 R^5 e^2 \frac{1}{\mu Q} \tag{8.10}$$

where n is the moon's mean motion about the planet, μ is the moon's rigidity ($6.5 \times 10^{11} \, \text{dyn cm}^{-2}$ for Io), e the eccentricity of the moon from which the tidal surface heating of Io can be calculated $\sim 41 \, \text{erg cm}^{-2} \, \text{s}^{-1}$.

All these considerations led to the conclusion that a $0.12 \, M_\oplus$ moon, $M_\oplus$ being the mass of Earth, in an Io-like orbital resonance and possessing a Ganymede-like magnetic field could remain habitable for several billions of years. Therefore, systems belonging to *47 UMa* and *16 Cyg B* could have possible moons about their giant planets as candidates for extraterrestrial life.

8.2.5 Tidal Heating: Io

Io is one of the four Galilean satellites of Jupiter that were first seen by Galilei in 1609.

On the surface of Io 400 active volcanoes and more than 150 mountains have been detected as well as volcanic plumes extending up to 500 km. Tidal forces of Jupiter on Io are 6000 times stronger than those of the Moon on Earth. There are also tidal forces caused by the two other Galilean satellites *Europa* and *Ganymede* (these are comparable to the tidal force of the Moon). Furthermore, the strength of these forces varies because the orbit of Io is elliptical. The variation of the tidal forces of Jupiter due to Io's elliptical orbit is 1000 times the strength of the tidal forces of the Moon.

Summarizing, Io is perhaps the most geological active object in the Solar System because of

- Tidal heating from Jupiter

- Heating effects due to resonances with other satellites of Jupiter.

On Earth, the tidal force causes deformations of the whole Earth's crust between 20 and 30 cm. On Io, these deformations can reach up to 300 m.

---

[1]see also Habitability and Cosmic Catastrophes, A. Hanslmeier, Springer, 2009

8.3   Galactic Habitable Zone

The chemical and physical properties of stars and interstellar matter in a galaxy depend on the distance to the galactic centre. As life is based on elements heavier than helium and as its formation takes a considerable time, a habitable zone in a galaxy can be identified where these basic requirements for life are fulfilled.

8.3.1   Stellar Populations

A stellar population is a group of stars that resemble each other in spatial distribution, chemical composition, and age. There are different stellar populations in galaxies found on different locations. Note that in astrophysics all elements heavier than He are called metals. A typical spiral galaxy like our Milky way consists of:

- Thin disk: O and B stars, T Tauri stars. Relatively young stars, age between several million and 5 billion years. The distance from the galactic plane within which half of the population can be found is 250 pc, metals: 11. This value denotes the metal content with respect to the Sun (metal content 1). The older the population, the lower will be the content of metals.

- Thick disk: G and K stars, planetary nebulae. Age between 5 and 13 Gyr, distance from galactic plane 700 pc, metals 1/3.

- Halo: globular clusters, RR Lyrae stars. Age more than 11 Gyr, height 2500 pc, metals 1/30.

Stars in a globular cluster belong to the oldest stars in the Galaxy, being older than 10^{10}yr. About 140 globular clusters have been found in the Milky Way (example Fig. 8.3). The stars of these objects are most probably no suitable host stars for planetary systems because they are metal poor objects, restricting planets form out of their nebula resulting from their final evolution.

We see that older stars are up to 30 times poorer in metals than the Sun. The height above the galactic plane increases as the age of the population increases. The older the population, the less it is concentrated to the galactic plane. This gives important hints to the formation of a galaxy.

8.3.2   Definition of the Galactic Habitable Zone

In a typical galaxy there exists a metal gradient. Near the center, the metal content becomes higher. The star density is higher there, with more stars evolving into supernovae and the interstellar matter becomes enriched by metals.

A planetary system must therefore be located close enough to the galactic center so that a sufficiently high level of metals could exist. Only under these conditions rocky planets could have formed. The necessity of heavier elements such as carbon in forming of complex molecules for life is also evident.

Although planetary systems can only evolve within a certain distance from the galactic center they should not be too close to the galactic center risking perturbations by passing stars. Such perturbations will trigger hazards from comets moving into the inner planetary systems. Near the galactic center outbursts from supernovae and from the supermassive black hole at the center could cause strongly enhanced short-wavelength radiation bursts destroying the complex molecules needed for life.

Figure 8.3: Globular Cluster M55 from CFHT. The cluster has number 55 in Messier's catalogue and consists of about 100 000 stars. The distance from Earth is about 20 000 Ly. Credit: Jean-Charles Cuillandre (CFHT) & Giovanni Anselmi (Coelum Astronomia), Hawaiian Starlight.

Figure 8.4: The galactic habitable zone in a typical spiral galaxy.

Observations of extrasolar planetary systems seem to show a trend that when the metallicity of the star becomes too high, more massive planets start orbiting close to the stars those can destroy Earth-sized objects there.

It was shown by [102] that in a barred galaxy, the GHZ (see Fig. 8.4) will be more complicated to define. The central bar can change stellar orbits. This is also valid for our Milky Way Galaxy, since it is a barred galaxy. For the dynamics of such objects two effects are important:

- Stars in the bar: their motion describes complicated orbits.

- Stars outside the bar: the bars will also influence the orbits of stars in the whole galaxy. Stars passing close to the bar can either gain or lose angular momentum, due to a positive or negative torque by the bar. Some stars will therefore be captured by the bar while some stars eventually may reach the escape velocity from the galaxy.

8.3.3   The Evolving Galactic Habitable Zone

In the early stages of galaxy evolution the heavy elements to form terrestrial planets were only present near the center of the galaxy because there the concentration of stars was largest and some stars already had evolved and became supernovae that enriched the interstellar medium with heavier elements there. However, this was not a safe environment and perhaps no continuous galactic habitable zone existed.

Gradually, the heavy elements spread through the galaxy and terrestrial planets formed at greater and safer distances from the galactic center. The habitable zone appeared about 8 billion years ago at a certain distance from the galactic center. An annular GHZ formed. In our Galaxy, the galactic habitable zone is an annular region 7-9 kpc^2 from the galactic center. This zone becomes larger with time and is composed of stars formed between 8 and 4 billion years ago. This was described by [65].

75 % of the stars in the galactic habitable zone are older than the Sun. Their age is 1 Gyr older than the Earth. Therefore, our civilization might belong to the "young generation" of galactic civilizations ([64]).

8.4   Potentially Habitable Bodies

As we have shown, the definition of a habitable zone depends whether we consider a planetary system or a whole galaxy. Therefore, it seems easier to define habitable bodies. According to [58] four classes can be defined.

8.4.1   Classes of Habitable Bodies

- Class I: bodies on which stellar and geophysical conditions allow Earth-analog planets to evolve so that complex multi-cellular life forms may originate;

- Class II: bodies on which life may evolve but due to stellar and geophysical conditions that are different from the class I habitats, the planets rather evolve toward Venus- or Mars-type worlds where complex life-forms may not develop;

- Class III: bodies where subsurface water oceans exist which interact directly with a silicate-rich core, like Europa;

---

^{2}pc denotes parsec, 1 pc = 3,26 light years or 32,6×10^{12} km

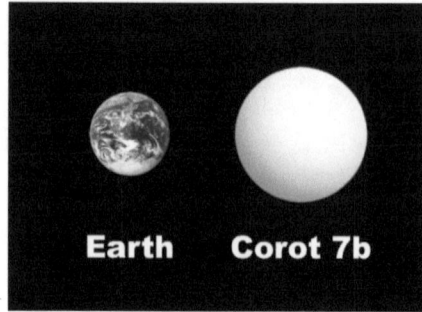

Figure 8.5: A size comparison of Corot 7b and Earth. Source: wikimedia.

- Class IV: bodies in which there are large liquid water layers between two ice layers or liquids above ice (also called super-Ganymedes or ocean planets).

8.4.2   Distribution of Habitable Bodies

The distribution of habitable bodies is at present difficult to estimate. Class I objects, Earth-analogues are at the onset of detection. Several Super Earth objects have been found already. An example is given in Fig. 8.5. Corot 7b has a diameter 1.53 that of Earth. It orbits the host star in only 20 hours and is very close to it (about 1/20 the distance between Mercury and Sun).

For example [1] report on a planetary system around the nearby M dwarf GJ 667C with at least one super-Earth in its habitable zone. GJ 1214b, the 6.55 Earth-mass transiting planet recently discovered by the MEarth team, has a mean density of ∼35% of that of the Earth. It is thought that this planet is either a mini-Neptune, consisting of a rocky core with a thick, hydrogen-rich atmosphere, or a planet with a composition dominated by water [29]. Habitability of super-Earth planets around main-sequence stars including red giant branch evolution was studied by [28].

As it was pointed out already, for a planet to be habitable also several protection mechanisms must be provided. One of these is an intrinsic magnetic field based on a dynamo. Magnetic fields of planets (like stars) are related to their rotation. A magnetic field of long duration depends mainly on two factors:

- planetary mass,

- rotation rate.

It seems that low-mass Super Earths ($M \leq 2\,M_E$) develop intense surface magnetic fields but their lifetimes will be limited to 2-4 Gyrs for rotational periods larger than 1-4 days ([112]).

Calculate the tidal heating effect of Earth on Moon and Moon on Earth?

Consider a hypothetical Jupiter at the orbit of Mercury in our Solar System. What would be the tidal heating on it?

Activities

In recent years many hot Jupiters have been detected. Imagine a hot Jupiter near a central star of one solar mass. Calculate:

- tidal locking
- tidal heating

Could there exist habitable satellites around a hot Jupiter?

Send Orders of Reprints at bspsaif@emirates.net.ae

CHAPTER 9

Origin of Elements

Abstract: In this chapter we discuss how, when, and during which processes the chemical elements have been formed in the universe. A short review about the origin of the universe is given The most abundant chemical element in the universe is hydrogen. This was formed during the first phases of the Big Bang and all other elements such as He, Carbon, Oxygen were formed by thermonuclear reactions (expect elements heavier than Fe). These processes are discussed in the chapter on stellar evolution where a distinction has to be done between low mass and massive stars. Therefore, it took several generations of stars before the universe contained a certain amount of these elements so that solid planets and complex molecules could originate.

The main message of this chapter is that the chemical elements necessary for life are the products of thermonuclear reaction that took place in the center of stars (with few exceptions), so that we consist of stellar dust.

Keywords: Universe: Origin; Primordial composition; stellar evolution; white dwarfs; red giants; neutron stars; black holes; supernovae; origin of elements; nucleosyntheis

Activities

Repeat which elements are important for life. Discuss the most fundamental chemical requirements for life:

- which elements

- physical environment

- which molecules

- microbiological aspects

9.1 The Primordial Composition

In this section we discuss how the most abundant elements in the universe, hydrogen and helium were formed immediately after the Big Bang. These elements constitute the primordial composition.

9.1.1   The Big Bang

The general concept about the origin and formation of universe has been summarized as the Big Bang Theory. There are several observational hints that the universe has a finite age and that it was formed from a state of extreme compactness and density about 13.6 Billion years ago.

- Hubble's law: E. Hubble (1889-1953) was the first who showed that besides our Galaxy the universe consists of many other galaxies. He was able to measure the distance of the Andromeda galaxy (using Cepheid variables, a relation between their intrinsic luminosity, absolute magnitude [1] and their variability time scale was measured) which is about the factor of 10 that of the diameter of our Galaxy[2]. But when he measured carefully the distances to several galaxies plotting the distances *versus* the measured radial velocities, he noticed that there exists a relation between the distance of a galaxy, d and its velocity of recession, v:

$$v = dH \tag{9.1}$$

H being the Hubble constant has a modern value of

$$H = 69.7 \pm 4.9 \, \text{kms}^{-1} \text{Mpc}^{-1} \tag{9.2}$$

That means: a galaxy at a distance of 1 Mpc (1 Mpc = 1 000 000 pc = $3,26 \times 10^6$ Ly) is receding from us with a velocity of about 70 km/s. The redshift z is given by:

$$z = \Delta\lambda/\lambda = v/c \tag{9.3}$$

$\Delta\lambda$ is the measured wavelength of a galaxy, λ is the wavelength at zero velocity, c is the speed of light and v the velocity. A galaxy that is moving away from us has a redshift according to the Doppler effect. According to Eq. 9.1 all galaxies seem to move away from us, the more distant they move, the larger becomes the velocity. H is Hubble's constant and by dimensional analysis we see that $1/H$ has the dimension of time. It gives an upper limit of the age of the universe (under the assumption of uniform expansion). This value is about 13.6 $\times 10^9$ y.

The expansion of the universe is shown in Fig. 9.1.

Calculate the distance of a galaxy with a redshift z=0.1.

- Cosmic Background Radiation, CMBR: G. Gamow and other astronomers predicted an isotropic radiation in the 1940s, that might have survived from the early hot universe. In 1964, Penzias and Wilson detected an isotropic radiation corresponding to a black body spectrum of an object at 2.7 K. This radiation is considered as the remnant since the universe has cooled to about 3000 K. At that time the recombination occurred: Before recombination, the free electrons scattered and absorbed radiation, as a result the universe became opaque. At about 3000 K atoms formed and the universe became transparent. This occurred when the universe was about 400 000 years old. The cosmic background radiation has a redshift of about 1000 making it observable in the microwave.

---

[1]see chapter 7.1.2
[2]Modern values: diameter of our Galaxy: 100 000 Ly, distance of Andromeda galaxy: 2.5 Million Ly

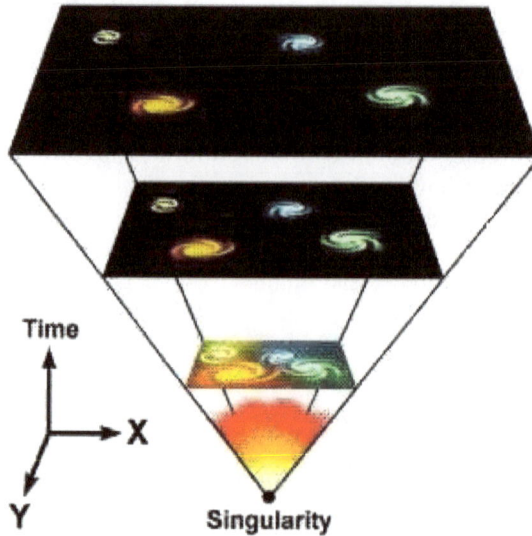

Figure 9.1: The expansion of the universe.

At what wavelength the CMBR can be observed? What was the original wavelength (using a redshift of z=1000)?

- Primordial nuclear fusion: By examining the spectra of very old stars, the primordial abundances of elements and isotopes can be measured. ^{4}He makes up to 24% of primordial matter, ^{3}He makes up one part in 10^4, deuterium one part in 10^5 and ^{7}Li one part in 10^9. The density required to produce these abundances is about $2...5 \times 10^{-28}\,\text{kg/m}^3$. The time for these nuclear reaction was about 200 s after time zero. Primordial fusion occurred during the first 3 minutes of the evolution of the universe.

These three observational facts are the background for the Big Bang theory describing the formation of the universe.

Discuss why only the two elements H and He were formed during primordial nuclear fusion.

9.1.2   Re Ionization

At a redshift of about 1000 and at an age of 400 000 years, the universe became cool enough for the start of the formation of atoms. In the process electrons combined with protons to form neutral hydrogen and the universe became transparent. Then the dark ages started because there were no light sources.

The second phase change occurred once objects started to form in the early universe energetic enough to ionize neutral hydrogen. As these objects formed and radiated energy, the universe went from being neutral back to being an ionized plasma, between 150 million and one billion years after the Big Bang (at a redshift $6 < z < 20$). By now, however, matter has been diluted by the expansion of the universe, and scattering interactions are much less frequent than those before recombination. Thus a universe full of low density ionized hydrogen has remained transparent, as is the case today.

Quasars are considered suitable objects to verify this theory. Although they are quasistellar but since they emit so much energy they are thought to be nuclei of very active galaxies. Some quasars are even detectable as far back as the epoch of reionization.

The first cosmic structure formation (see Fig. 9.2) is reviewed by [23]. The evolution and observational aspects of the first galaxies are described by [12], while early star-forming galaxies and the reionization of the Universe has been described in [85].

9.2 Evolution of Stars

The evolution of stars strongly depends on their mass. The larger the mass the faster will be the evolution.

9.2.1 Low Massive Stars

Low mass stars are found on the right part of the Hertzsprung-Russell diagram. The lower the mass, the larger will be the main sequence lifetime. They are thought to be the ideal host stars for planetary systems with habitable zones. Our Sun being a G2 V star has a total lifetime of about 9 billion years on the main sequence. For stars like the Sun or less massive, the hydrogen fusion is the main energy source, while for stars more massive than the Sun the Carbon cycle acts as the main energy source. Before the nuclear fusion starts, or any time a star or part of it shrinks in size, gravitational energy is released. This occurs particularly in the pre main sequence evolution. After a fuel has been used up, a star also contracts and the energy released during such a contraction heats the core. The gravitational energy is given by:

$$E = \frac{GM^2}{R} \tag{9.4}$$

If we calculate the gravitational energy for the Sun then:

$$E = \frac{6.67 \times 10^{-11}(2 \times 10^{30})^2}{7 \times 10^8} = 4 \times 10^{41} \text{ J} \tag{9.5}$$

This means that the Sun has enough energy for 30 million years[3] by contraction assuming it radiates the present luminosity.

The luminosity of a star changes during its evolution and therefore the location in the HR diagram. This is also called evolutionary track.

In Fig. 9.3 the HR diagram for stellar cluster at different ages is given. Stars form from a fragmentation of large interstellar clouds, therefore they are born in clusters of stars and all of these stars have a bout the same age. At the beginning of stellar evolution the stars appear at the main sequence. Before, they evolve from the right part of the HR diagram to the main sequence, a main sequence star is characterized by the fact that of being in hydrostatic equilibrium, the nuclear fusion (hydrogen into He, or carbon cycle) works at its center. As soon as the hydrogen in the center is consumed up, the star undergoes a transformation. In

---

[3]This is also called Kelvin Helmholtz time

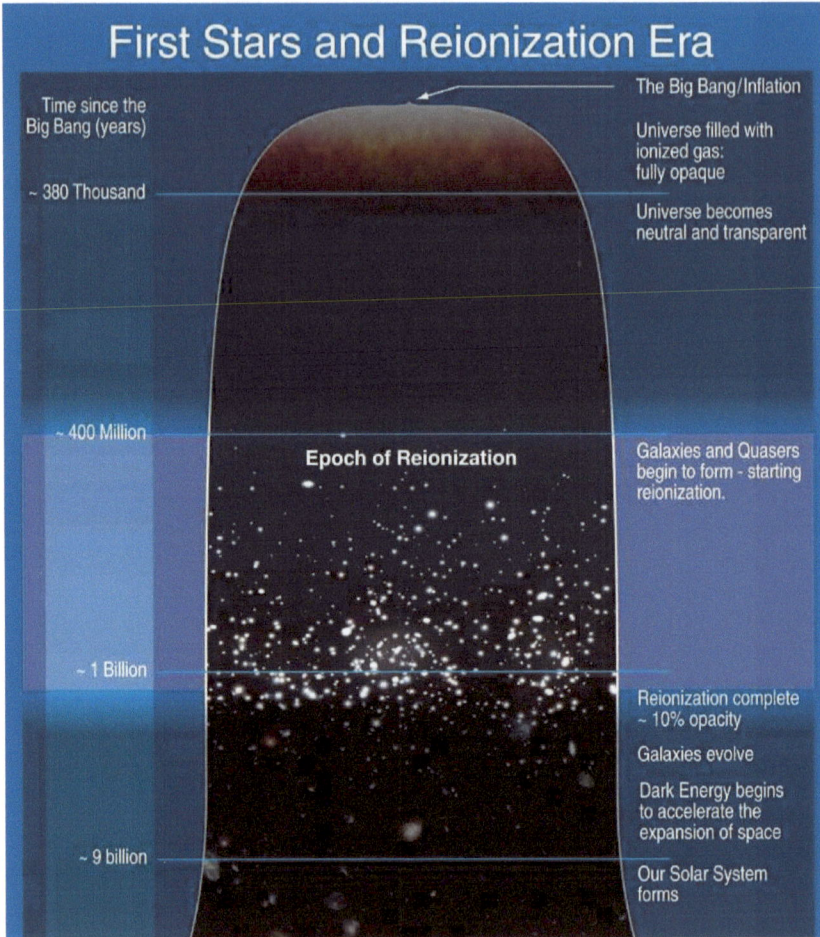

Figure 9.2: The evolution of the universe.

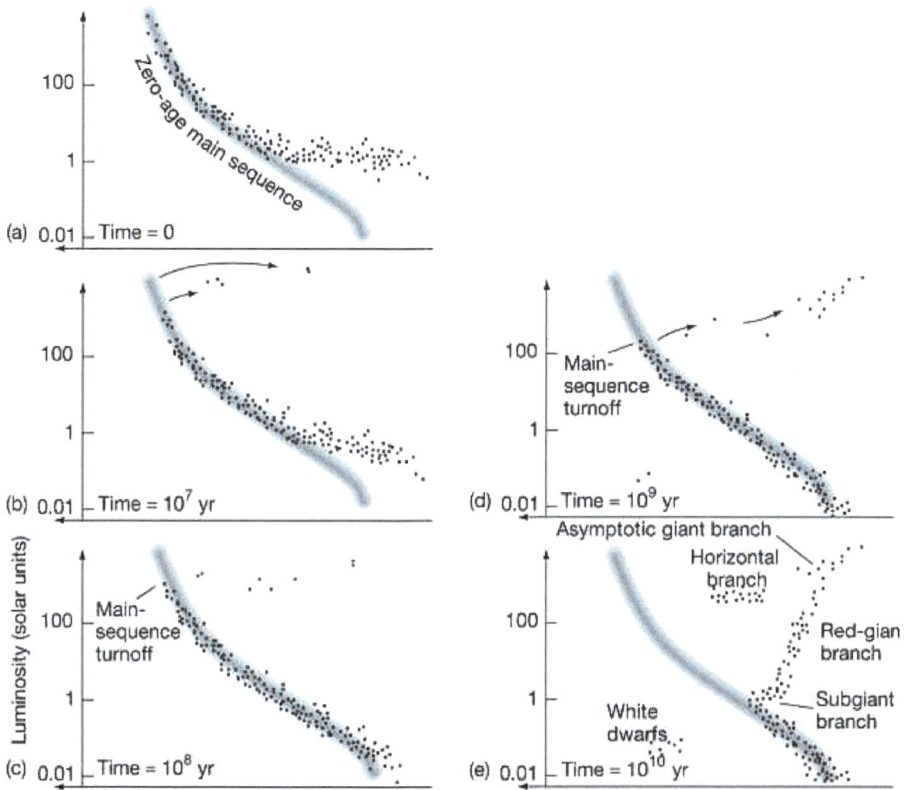

Figure 9.3: The HRD for different ages of stellar clusters.

the center no longer fusion reactions work, however, they work in a shell about the center. When the nuclear burning shell slowly progresses outwards, the center starts to shrink. As we have seen above, a shrinking means a release of gravitational energy, that heats the core. The star as a whole expands to a red giant, increasing its diameter by more than 100 times. When the Sun becomes a red giant, it expands beyond the Earth's orbit. At some point the temperature in the core becomes high enough so as to initiate other nuclear burning reactions, like the triple Alpha process where $3^4\text{He} \rightarrow^{12} \text{C}$. At this stage the star reaches the highest point in the HR-diagram. It maintains nearly the same temperature as it had as a main sequence star but its luminosity increases several has increased its orders of magnitude, which is easily explained: the surface of a star has increased and the luminosity depends on:

$$L = 4\pi R^2 \sigma T^4 \tag{9.6}$$

The star has now two sources of energy production:

- one at the core (He $\rightarrow$ C),

- one at a shell (H $\rightarrow$ He).

$\rightarrow$It becomes unstable.

Depending on the mass of the star, further nuclear fusion processes start, therefore all elements like C, O, N,... up to the element Fe are formed. A production of elements heavier than Fe requires more energy than it releases, thus being very rare it occurs only in very energetic processes like during a supernova explosion. The ultimate fate of a low mass star is a white dwarf, the star loses its outer shell and the remnant becomes a compact object about the size of Earth. This will also be the ultimate fate of the Sun.

Summarizing the evolution of low mass stars as follows:

- Evolution towards the main sequence, protostar, T Tauri phase; the luminosity of the star comes from release of gravitational energy; since the star is unstable, no circumstellar habitable zone can exist.

- Main sequence evolution: the star remains relatively stable at the main sequence in the HR-diagram; for a star with 1 solar mass this lasts for about 10^{10} years. The luminosity of a star slightly increases; for the Sun this increase was about 30 % over the past 4.5 billion years. At this stage a stable circumstellar habitable zone could exist.

- Evolution toward red giant. This starts when the nuclear fuel at the core is consumed up; the red giant phase lasts only for several 10^8 years. When former habitable zones about a star become destroyed, they quickly progress outwards.

- Final evolutionary state: for stars with masses less than 1.4 solar masses (Chandrasekhar limit) the final stage is a white dwarf. These compact Earth-sized objects slowly cool and last in principle forever, where no further nuclear reactions occur. A habitable zone around a white dwarf may exist in principle but has to be very close to the star because of the low luminosity.

Consider the evolution of the Sun towards a red giant. How will the habitable zone evolve in the Solar System?

9.2.2   Massive Stars

Stars with masses larger than $1.4\,M_\odot$ evolve differently especially their final stages. The Chandrasekhar limit is specified as the maximum mass of a stable white dwarf star[4]. Above about 1.4 solar masses, electron degeneracy pressure in the core of a star becomes insufficient to balance the gravity of the star. Electron degeneracy is a quantum-mechanical effect. It can be understood from the Pauli Exclusion Principle. Since electrons are fermions, no two electrons can exist in the same state. The equation of state for an ideal gas is:

$$P = NkT \tag{9.7}$$

For electron degeneracy the pressure becomes:

$$P = K_1 \rho^{5/3} \tag{9.8}$$

for the non relativistic case or

$$P = K_2 \rho^{4/3} \tag{9.9}$$

for the relativistic case. K_1, K_2 are constants, ρ is the density.

As soon as $1.4\ M_\odot$ mass limit is reached, the object collapses. This can be observed as a supernova explosion. There exist two types of supernovae:

- Type II: when the core of a massive star exceeds the Chandrasekhar limit it collapses and after the implosion of the core a shockwave moves outwards and the outer shell of the star is blown off. This can be observed as a strong increase in luminosity of a star. A supernova is as luminous as a whole galaxy.

- Type I: consider a double star where one component is a white dwarf. When this white dwarf accretes matter from a companion its mass exceeds the Chandrasekhar limit and a supernova type I can be observed.

In both cases supernovae are extremely bright and occur when the final core or the total mass of a star exceeds the Chandrasekhar limit, therefore all supernovae have about the same luminosity and by comparing their brightness with the measured brightness (which depends on their distance), we can easily calculate their distances. An example of such a supernova explosion is that which occurred at a distance of about 6300 Ly in the year 1054. The exploding star was visible with the naked eye even during daytime as was reported by Chinese astronomers. The Crab Nebula (Fig. 9.4) thought to be the remnant of this explosion.

During the explosion the star can become as bright as a whole galaxy (absolute magnitude about -18^M, see Fig. 9.5).

The ultimate fate of a massive star is a neutron star (for masses less than 3 to 4 solar masses) or a black hole. Neutron stars are extremely compact, about the size of 10 km, they form when the electrons and protons combine to generate neutrons. The pressure of degenerate neutrons prevents further contraction. Black holes form when the pressure of degenerate neutrons is not sufficient to prevent a final collapse. The radius of a black hole can easily be estimated from even classical physics. The velocity of escape v from a body with mass M and radius R is:

$$v = \sqrt{\frac{2GM}{R}} \tag{9.10}$$

---

[4]S. Chandrasekhar, 1930

Figure 9.4: The Crab nebula is the remnant of a supernova explosion that was observed by the Chinese astronomers in 1054; it is at a distance of 6500 Ly.Credit: NASA-HST .

Figure 9.5: Supernova explosion in a distant galaxy. Credit: NASA-HST.

If we assume that $v \to c$, then the 'radius' of a black hole becomes:

$$R_{\mathrm{BH}} = \frac{2GM}{c^2} \tag{9.11}$$

Give two HRD (i) for a very young stellar cluster, (ii) for an old cluster.

9.2.3   Supernovae Explosions and Nucleosyntheis

Stars with masses larger than 1.4 solar masses explode in a supernova. There are more than hundred known chemical elements that exist naturally in the universe. Stellar evolution explains the relative abundances of the elements and their isotopes. Elements like hydrogen and He as well as carbon and oxygen are produced by nuclear fusion reactions occurring near the stellar cores. The fusion of carbon and oxygen results into other elements like silicon, sulfur and magnesium. The fusion of heavier elements requires higher energies and therefore temperatures since the repulsive Coulomb forces become stronger for heavier nuclei.

When the core temperature is about 1 billion K the radiation consists of extremely high energetic gamma rays. These are able to disrupt nuclei causing them to emit protons, neutrons and Alpha particles. These particles immediately react with other nuclei. Fusion of an element with a mass number[5] greater than 60 requires more energy than it produces. Elements heavier than Fe and Ni are relatively rare but they exist. Moreover, they are extremely important for life (*e.g.* copper, gold, zirconium, iodine,...). These elements are not produced by fusion but by reactions between nuclei and neutrons. There are two different processes:

- s-process: since neutrons are electrically neutral, they can be easily captured by other massive nuclei. The buildup of massive nuclei by neutron capture starts during the red giant phase. The rate of the neutron capture is slow compared to the decay rate, therefore this process is called s-process (s is for slow). Elements up to bismuth (83 protons, 126 neutrons) can be built up.

- r-process: very heavy or neutron-rich nuclei can be produced only when there are so many neutrons that neutron capture is very rapid (r) compared with decay rate. The required high neutron density occurs only during the collapse phase of the core when a supernova explodes.

Supernovae are important for two reasons for the evolution of life:

- during the core collapse heavy elements are produced,

- heavy elements are propelled into space during the explosion.

Our Sun formed from gas enriched with heavier elements that were formed during supernova explosions of stars of former generations. The universe becomes slowly enriched with elements heavier than He. These elements are processed in the stellar cores or during a supernova explosion. The cosmic abundance of elements is given in Fig. 9.6

---

[5]The mass of protons and neutrons

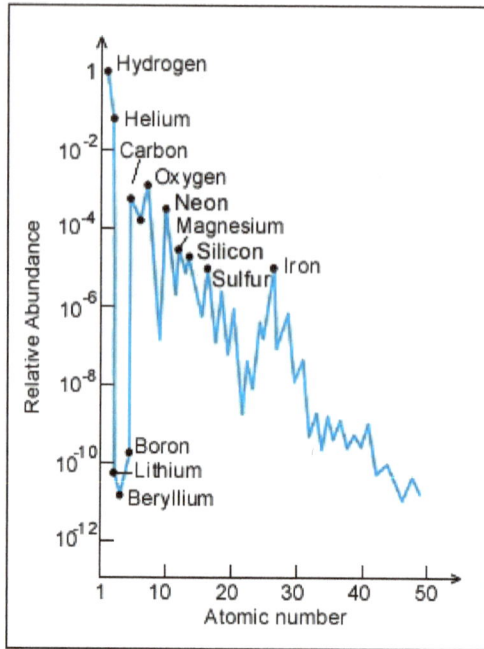

Figure 9.6: Cosmic abundance of elements.Credit: www.fas.org/irp

Activities

In the constellation of Taurus there is the open star cluster Hyades. Try to get data of stars belonging to that cluster. Use the data to construct a HRD of the Hyades cluster. Interpret the results.

- How old is the cluster?

- What is the proportion of main sequence stars to the whole stellar population in the Hyades cluster?

- Which stars would be interesting for astrobiology?

- Select 10 target stars for future astrobiology projects.

- Could life have evolved on hypothetical planets there?

Send Orders of Reprints at bspsaif@emirates.net.ae

In Situ and Remote Sensing for Life

Abstract: In this chapter we discuss the several attempts to search for life in the universe. Parts of this chapter are extracted from chapter 10 of another book of the author [41]. The Drake equation which estimates the probability for extraterrestrial civilizations being on the same technological level as we is discussed and approvements are suggested.

- Active search: this comprises *in situ* measurements. Several objects, like Venus, Mars, Titan and some asteroids in the Solar System have been visited by landers. One of the main aims of these landers was the to search for any biologic activity by measuring the surface conditions on these objects. Some space missions conducted to explore the Solar System carry messages about our civilization. Of course the chance of finding an extraterrestrial civilization through these space vehicles is extremely low.

- Passive search: this comprises all methods to detect (i) planets, (ii) speculate about satellites of these planets, (iii) find spectroscopic signatures of biologic activity (biomarkers) such as oxygen, ozone, methane and other compounds in a planetary atmosphere.

. Examples of radio messages sent to different objects are given. No answer has been received so far.

The main message of this chapter is that from several arguments we can conclude that our civilization might not be the only one in the Galaxy and that we should belong to the older generation of civilizations at least in the Galaxy.

Keywords: Extraterrestrial life; biomarkers; Drake Equation; Moon; Mars; Venus; Europa; Titan; Mars: rovers; SETI; interstellar communication; radio messages

10.1   Active Search

10.1.1   Venus

Although, Venus was visited by several space missions but only the Russian Venera mission succeeded in landing on its surface. Since found to be extremely hostile to life no specific measurements for any life signature were made.

The Soviet Venus exploration program extended over more than two decades with an ambitious aim to land on Venus but already in the 1960s it was thought that the surface temperature of Venus was approximately 300^0C, with an atmosphere consisting mainly of carbon dioxide and nitrogen at about 20 bars. Consequently, the capsule was designed to survive 300^0C and 25 bars. Venera 4, launched in 1967, reached Venus and transmitted

Figure 10.1: Model of Venera 13. Source: JPL.

data on the atmosphere and environment until it reached an altitude of about 26 km after 93 minutes of descent.

Venera 9 and 10 transmitted data from the surface for 53 and 65 minutes, respectively. The landers capabilities were not the limiting factors in the surface survival time; instead, each mission had to be terminated when its orbiter exited the communication range. Venera 11 was launched on 9 September 1978 followed by Venera 12 that was launched on 14 September 1978. Both traveled to Venus with a separation of about 0.02 AU and this separation increased to 0.5 AU. The Venera spacecraft could either be spin stabilized with a spin period of several hours, or 3 axis stabilized. The Venera 11 mission was terminated in February 1980 and the Venera 12 mission ended in April 1980. The next series of landers on Venera missions 11, 12, 13, and 14, (all launched in 1978 within a difference of few days from each other) improved upon the successes of the ongoing Soviet Venus exploration program. These missions all descended to the surface in approximately one hour and lasted on the surface for up to two hours. Venera 11 transmitted data for about 95 minutes, until the flyby spacecraft used as relays went out of reach. Venera 12 transmitted data for 110 min. Unfortunately, not all the experiments on Venera 11 and 12 succeeded. In Fig. 10.1 a model of Venera 13 is shown. In Fig. 10.2 an image showing the surrounding surface of the landing site is given.

Landing on Venus, the past and future is described in [4].

10.1.2 Moon

The most ambitious space mission made so far was to bring astronauts to the surface of the Moon. In the Apollo 11 mission astronauts Neil Armstrong and Buzz Aldrin (Fig. 10.3) landed their Lunar Module (LM) on the Moon on July 20, 1969 and walked on its surface while Michael Collins remained in lunar orbit in the command spacecraft, and all three landed safely on Earth on July 24. Five subsequent Apollo missions also landed astronauts on the Moon, the last in December 1972. In these six spaceflights, 12 men walked on the Moon.

The concept to reach the Moon, land there and return to Earth was realized by the following steps:

- The Command Module (CM) was the crew cabin, surrounded by a conical re-entry

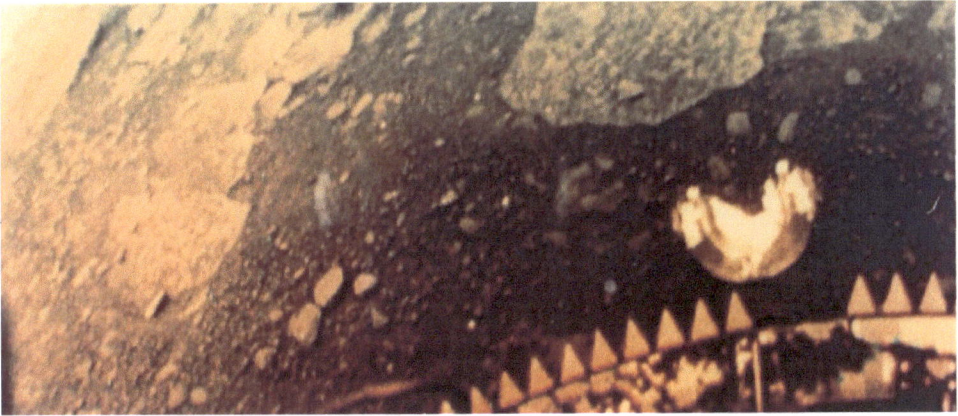

Figure 10.2: Surface of Venus at Venera 13 landing site. Source: IKI.

Figure 10.3: The second man on the Moon, Ed. Aldrin photographed by N. Armstrong. Credit: NASA

Figure 10.4: The Apollo 16 Landing Module. Credit: NASA

heat shield, designed to carry three astronauts from launch to lunar orbit and back to an Earth ocean splashdown.

- Lunar module, LM (Fig. 10.4): designed to fly between lunar orbit and the surface, landing two astronauts on the Moon and taking them back to the Command Module. It had no aerodynamic heat shield and was of a construction so lightweight that it would not have been able to fly through the Earth's atmosphere. It consisted of two stages, a descent and an ascent stage. The descent stage contained compartments carrying cargo such as the Apollo Lunar Surface Experiment Package and Lunar Rover.

Of course, no biologic activity was expected at the lunar surface. On the other hand, the Moon offers the possibility of unravelling a better understanding of the conditions for habitability on the Earth and the along with conditions for life on the early Earth. Several tests can be performed there:

- the linearity or non-linearity effects of different magnitudes of space environmental stresses on organisms, particularly gravity;

- cumulative environmental effects both in individual organisms and across generations,

- the synergistic effects of different space environmental parameters on organisms.

The close proximity and scientific importance of the Moon make it a useful permanent location and staging post for the human expansion into space ([26]).

10.1.3   Mars

The first search for biomarkers on the Martian surface was made by the *Viking 1* and *Viking 2* landers. Viking 1 Lander touched down in the Chryse Planitia and transmitted the first image was 25 seconds after landing. The lander operated for 2245 sols until November 13th,

1982 [1]. Viking 2 lander touched down on the Martian surface in the Utopia Planitia on September 3, 1977. Due to a battery failure, it stopped transmission on April 11, 1980 after 1281 sols. There were four biology experiments.

- Gas chromatograph-mass spectrometer (GCMS): vapor components can be measured and their molecular weight of each chemical can be determined. Untreated Martian soil was partly heated partly analyzed in original state but no significant amount of organic molecules was found.

- Gas exchange, GEX: a Martian sample was incubated with organic and inorganic nutrients. Also water was added in a later stage. Using a gas chromatograph, it was tested whether the concentrations of gases like O, CO_2, N, H and CH_4 changes.

- Labelled Release LR: a sample of Martian soil was inoculated with a drop of very dilute aqueous nutrient solution (containing the Miller-Urey products).

- Pyrolytic Release, PR: in this experiment it was tested whether some ^{14}C marked biomass can be found. An artificial ^{14}C rich Martian atmosphere was provided and if there would be some respiration process then part of this ^{14}C should be found in the biomass.

Other successful Mars exploration rovers are *Spirit* (landed on Jan 3, 2004 in the Gusev Crater, once believed to have been a crater lake) and *Opportunity* (landed on Jan 24, 2004 in Meridiani Planum, where there are large deposits of hematite, indicating the presence of past water).

The *Astrobiology Field Laboratory* (AFL) as an unmanned spacecraft to explore Mars and would be launched around 2016.

Summarizing the results of the biology experiments made *in situ* on Mars, we can state the analyzes by gas chromatography and mass spectrometry did not show the presence of any organics above the ppb level in the top few centimeters of the martian soil (Biemann et al 1977 [10]). We have to mention however, that the Viking mission also indicated that there is a ubiquitous layer of highly oxidizing aeolian material covering the surface. This layer could have oxidized organic material and is therefore responsible for the lack of detection of organic matter on the surface of Mars. This planet may therefore be self-sterilizing (Mancinelli, 1989 [69]). Organic degradation under simulated Martian conditions was tested by Stoker and Bullock, 1997 [100]. They used chambers containing Mars-analog soil mixed with the amino acid glycine, evacuated and filled to 100 mbar pressure with a Martian atmosphere gas mixture and then irradiated with a broad spectrum Xe lamp. They found that organic compounds are destroyed on Mars at higher rates than the rate they are deposited by meteorites and no strong oxidants are needed to explain this destructive process.

In July 2012, the car sized Mars rover Curiosity successfully landed and will be able to move up to 200 m per day on the Martian surface.

10.1.4 Europa

Europa is one of the Galilean satellites of Jupiter. It is conceivable that life could exist there beneath a sub-surface ocean of liquid water. Moreover, undersea volcanic vents are highly probable. These are the conditions similar to those found on early Earth. At the surface a temperature of the ice crust of -180° C can be expected. The surface is exposed to

---

[1]The transmission broke down due to a faulty command sent by ground control.

Figure 10.5: The Huygens probe was built by the European Space Agency. It was flown to the US from Europe. ESA. Source: ESA.

various hazards such as intense radiation from Jupiter's magnetosphere, impact of comets and meteoroids, and solar UV radiation. It can be estimated that below about 10 cm these hazards become very small. At the same time, because of tidal heating (see previous chapter) the temperature increases with depth, the temperature gradient is estimated about $0.1°$ C/m. Therefore, in a depth of a few km the temperature becomes more than $0°$ C.

We have discussed already the threat to habitability by highlighting the impacts of asteroids and comets however they have also some positive effects. Impacting comets on Europa's icy crust might have provided the necessary organic compounds. This was investigated by Pierazzo and Chyba, 2002 [79].

The presence of an ice ocean under the surface for other Galilean satellites has already been discussed in the context of tidal heating.

10.1.5 Titan

Saturn's largest satellite Titan was visited by the Titan lander called Huygens Probe. The combined Cassini-Huygens spacecraft was launched from Earth on October 15, 1997. The lander called Huygens (Fig. 10.5) separated from the Cassini orbiter on December 25, 2004, and landed on Titan on January 14, 2005 near the Xanadu region. This was the first landing (Fig. ??) ever accomplished in the outer Solar System. Since there are lakes on Titan it was not sure whether the probe will land on solid surface or in a lake and a touch down in an ocean was also taken into account in its design. The probe was designed to gather data for a few hours in the atmosphere, and possibly a short time at the surface. It continued to send data for about 90 minutes after touchdown. It remains the most distant landing of any craft launched from Earth. The probe remained dormant throughout the 6.7-year interplanetary cruise, except for bi-annual health checks. The main mission phase was a parachute descent through Titan's atmosphere. The batteries and all other resources were sized for a Huygens mission for a time period of 153 minutes.

Several instruments were on board the Huygens lander. The Surface-Science Package (SSP) contained a number of sensors designed to determine the physical properties of Titan's surface at the point of impact, whether the surface was solid or liquid. An acoustic

Figure 10.6: Huygens *in situ* image from Titan's surface; the only image from the surface of a planetary body outside the inner Solar System. Source: ESA/NASA.

sounder, activated during the last 100 meters of the descent, continuously determined the distance to the surface, measuring the rate of descent and the surface roughness (*e.g.*, due to waves). The ACP experiment (Aerosol Collector and Pyrolyser) drew in aerosol particles from the atmosphere through filters, then heated the trapped samples in ovens (using the process of pyrolysis) to vaporize volatiles and decompose the complex organic materials. Other instruments were the Gas Chromatograph Mass Spectrometer (GC/MS), the Huygens Atmospheric Structure Instrument (HASI), and Doppler Wind Experiment (DWE).

A detailed image of the landing site is shown in Fig. **??**.

The surface of Titan has been revealed globally, if incompletely, by Cassini observations at infrared and radar wavelengths as well as locally by the instruments on the Huygens probe. Extended dune fields, lakes, mountainous terrain, dendritic erosion patterns and erosional remnants indicate dynamic surface processes ([45]). Results from the Huygens probe on Titan are discussed in ([60]).

10.1.6   Do we Contaminate the Solar System?

We have discussed that life can survive under extreme conditions. Is there a danger that by human space activity, the objects investigated might be contaminated?

There exist several UN Resolutions for Planetary Protection: 610 UNTS 205 (January 27, 1967) and A/RES/34/68 (December 05, 1979). All signatory states of the OST have ratified. E.g. NASA Policy Directive NPD 8020.7 and Planetary Protection Offices at NASA and ESA

Several categories have been defined:

- Category I: Missions to Moon, Venus and Dwarf/Minor Planets.

- Category II: Comets, Jupiter, Pluto/Charon, Kuiper Belt Objects *etc*

- Category III: Flybys for Mars and Europa (no direct contact)

- Category IV: Lander-Missions to Mars and Europa.

    - IV a: bioburden $< 3 \times 10^5$ total, < 300 spores/m^2
    - IV b: life detector-bearing hardware: < 30/m^2
    - IV c: anything with $0.5 <$ water activity $< 1, T > -25^0$ for > 500 yrs Special Regions = Planetary Parks, for at least 50 yrs

- Category V: Missions to Earth (*e.g.* Mars Sample Return)

10.2 Theoretical Estimations of Extraterrestrial Life

Several attempts have been made to estimate the number of civilizations in our Galaxy or even in the whole universe. These attempts are based on theoretical considerations and many uncertainties still remain because the input parameters are not known with sufficient accuracy. However, in the past 20 years great progress in determining these numbers has been made.

10.2.1 Fermi's Paradoxon

Fermi raised in the 1950s the following question: if there are thousands or millions of civilizations in our Galaxy, where are they, and why they have not visited us since ages, the paradoxon thus stating if the universe is full of civilizations then we we should have already been contacted or visited by some of them.

This discussion was made in the 1950s when UFOs (unidentified flying objects) were very popular. Fermi also argued that considering the age of the Galaxy and most stars in it, many civilizations must have risen before our own, so that we belong to the younger civilizations. As we have already seen, organic matter appeared after the disappearance of rapidly evolving first generation stars. It can be estimated how long it would take for advanced civilizations to colonize the whole Galaxy: the number is just serval million of years being not much compared to the age of the Galaxy[2]. There are several statements to solve this so called Fermi paradoxon:

- Interstellar travel is not possible and does not matter with the scientific and technological level reached by a civilization.

- Advanced civilizations have little or no interest in expanding through large regions of the galaxy.

- Technological civilizations annihilate themselves, or disappear by natural catastrophes, before having the chance to spread through large regions of the galaxy.

- Alien civilizations have not reached us yet because intelligent life is extremely difficult to emerge. Otherwise alien civilizations would necessarily be here. As a result we could find ourselves among the most evolved technological civilizations in our galaxy or we could even be the only one.

Others claim that maybe some Gods from heaven could have been in fact alien Astronauts or that we are too primitive to Aliens so that they don't want to communicate with us. Theoretical physicists think that the fact that we do not see aliens around could be the first

---

[2]Fermi gave values between 5 and 50 million years to colonize a whole galaxy.

Table 10.1: Parameters those have been used by *Drake* in his equation.

Variable	Value	Comment
R^*	10/yr	10 stars formed per year in the Galaxy
f_p	0.5	half of all stars have planets
n_e	2	two planets per star are habitable
f_l	1	100 % of the habitable planets form life
f_i	0.01	1 % of them have formed intelligent life
f_c	0.01	1 % of them are able to communicate
L	10 000 yr	lifetime of civilizations

proof of the existence of brane worlds: all advanced aliens would have emigrated to better parallel universes. Brane worlds mean that our Universe with three space dimensions is located in a subspace (brane) of a higher dimensional Cosmos. These models allow large, and even infinite, extra dimensions. For more discussions on these subjects see [24].

10.2.2   Drake Equation

Following discussion of *Fermi*, *Drake* (in 1960) tried to define an equation that gives the number of civilizations in our Galaxy we are able to communicate with:

$$N = R^* f_p n_e f_l f_i F_c L \qquad (10.1)$$

In this equation the following parameters enter:
R^* Rate of star formation in the Galaxy
f_p Fraction of those stars that have planets
n_e Average number of planets that are habitable per star that has planets
f_l Fraction of n_e that actually have developed life
f_i Fraction of f_l that have developed intelligent life
f_c Fraction of civilizations that have developed a technology similar to ours and that release signals
L Length of the time interval where such civilizations send out detectable signals into space (maybe also the lifetime of such a civilization).

In order to estimate N, Drake has used the values that are shown in Table 10.1.

The question is, how accurate can these values be determined today (see also Burchell 2006 [13])? The number of stars being formed in the Galaxy, R^* seems well known and very close to 10/yr. n_e is unknown. Here the new planet finding missions (GAIA, TPF and Darwin) will help to get better estimates. Up to now it seems more likely to find giant gas planets near stars that are definitely non habitable objects. However, as it has been mentioned already several times, habitability must be extended to other objects such as satellites of gas giants (*e.g.* Jupiter's satellite Europa). Thus n_e is a rather uncertain number.

The value f_l can be estimated from the Earth's history only. On Earth, life began as soon as the conditions became favorable. Abiogenesis describes the generation of life from non-living matter, such as the generation from non living but self-replicating molecules near hydrothermal vents *etc.* The value of f_l would be more optimistic if life independent from Earth would be detected on Mars or on Europa.

The next two factors in Drake's equation are still more uncertain: f_i, f_C. On Earth it lasted about 3.5 billion years from the first forms of life to the evolution of humans that are able to communicate. Thus alien civilizations would be expected on relatively old planets.

There is an important argument against all SETI [3] activities. If civilizations are dispersed both in space and time around a galaxy then the chance to interact would become extremely low. It seems however, that times required for biological evolution on habitable planets of a galaxy are highly correlated, *i.e.* planets in habitable zones evolve nearly at the same time in a galaxy, so that at least there does not seem a dispersion in time (Vukotic, Cirkotic, 2007 [107]).

10.2.3   A New Approach

Of all the main sequence stars about 8 % are of type G. If we assume that the Galaxy contains 200×10^9 stars then there would be 16 Billion G type stars. Let us further assume that 10 % of these stars are in the habitable zone of the Galaxy, therefore remaining 1.6 Billion stars. The occurrence of planetary systems seems quite common, so we can assume that 1/3 of these stars has planetary systems, thus about 0.5 billion planetary systems in the habitable zone of the Galaxy remain. Since the habitable zone around G stars is relatively broad, the probability to find planets there is quite high. However there should be an Earth sized planet with a large satellite that stabilizes its rotational axis. If the percentage of G stars that have planetary systems and a planet of Earth size in its habitable zone is 0.1 % then there remain still 500 000 objects.

Thus our crude estimate gives 500 000 Earth like planets in the habitable zone of the Galaxy that are around a G-type star and that lie in the habitable zone with stable axis of rotation. Even if we assume that from these 500 000 objects only 1/1000 contains a continuous habitable zone (no nearby supernova explosion, no catastrophic impacts...), still 500 object remain.

However, we must take into account the age of these 500 objects. As we know from the Earth, it took about 3.5 billion years for life to develop to higher forms. How many of these 500 objects are old enough? If we take a crude value of 1/10 then finally our very crude approximation gives 50 living civilizations in the Galaxy in the habitable zone being roughly at the same level than we are and which have survived cosmic catastrophes resp. while major catastrophes have occurred on these planets. Taking into account the enormous distances in the galaxies, we immediately see how difficult it would be to communicate with these civilizations. This estimation could explain why they have not been answered yet.

Finally, maybe the above made estimation is even too pessimistic. We have considered only G stars. Maybe even the cooler K stars could be the candidates for hosting planets in habitable zones. The percentage of these stars is 13 % of all main sequence stars. Civilizations orbiting a K-star could profit from a longer main sequence lifetime of their host star. If life would have been extinct there by one of the cosmic catastrophes described above and in the previous sections, then maybe there was a "second chance" or there will be one.

Taking K stars with the argument of a "second chance" for life into account this would add another say 100 civilizations to our Galaxy.

Finally, as it was discussed, some of the large satellites orbiting giant planets and even isolated free floating planets are possible habitable. Thus the number of habitable worlds strongly increases by some unknown factor.

While all these estimates are very crude but also very conservative we still may expect at least 100-200 civilizations being similar to our civilization in our Galaxy. This is not a large

---

[3]Search for Extraterrestrial Intelligence

number but definitely different from one, because at least to my opinion, the parameters to calculate the values have been taken quite conservative. Let us hope to find them!

10.3   Radio Observations

SETI is a programme for the Search of Extraterrestrial Intelligence. There are several attempts.

10.3.1   Radio Communication with ET

Already in 1896, N. Tesla proposed to use radio frequencies that penetrate the Earth's atmosphere to get in contact with extraterrestrial life. His experiments revealed some signals being quite different from Earth's noise and Tesla was sure that the signal came from Mars. On August 21-23, 1924, Mars entered an opposition closer to Earth than any time in a century before or since. In the United States, a "National Radio Silence Day" was promoted during a 36-hour period from the 21-23, with all radios quiet for five minutes on the hour, every hour. The aim was to find out whether the signals measured by Tesla were really of Martian origin. Frank Drake performed in the 1960s the first modern SETI experiment, named "Project Ozma", after the Queen of Oz in L. Frank Baum's fantasy books. Drake used a radio telescope 26 meters in diameter at Green Bank, West Virginia, to examine the stars Tau Ceti and Epsilon Eridani near the 1.420 gigahertz marker frequency. This is a famous region of the radio spectrum also called the "water hole" due to its proximity to the hydrogen and hydroxyl radical spectral lines. In 1980, Carl Sagan, Bruce Murray, and Louis Friedman founded the U.S. Planetary Society, partly as a vehicle for SETI studies.

In the early 1980s, Harvard University physicist Paul Horowitz took the next step and proposed the design of a spectrum analyzer specifically intended to search for SETI transmissions.

SETI@home was conceived by David Gedye along with Craig Kasnoff and is a popular volunteer distributed computing project that was launched by the University of California, Berkeley in May 1999. It was originally funded by The Planetary Society and Paramount Pictures, and later by the state of California. The project is run by director David P. Anderson and chief scientist Dan Werthimer. Any individual can become involved with SETI research by downloading the Berkeley Open Infrastructure for Network Computing (BOINC) software program, attaching to the SETI@home project, and allowing the program to run as a background process that uses idle computer power. The SETI@home program itself runs signal analysis on a "work unit" of data recorded from the central 2.5 MHz wide band of the SERENDIP IV instrument. After computation on the work unit is complete, the results are then automatically reported back to SETI@home servers at UC Berkeley. As of June 28, 2009 the SETI@home project has over 180,000 active participants volunteering a total of over 290,000 computers.

In 1974, a largely symbolic attempt was made at the Arecibo Observatory to send a message to other worlds. It was sent towards the globular star cluster M13, which is 25,000 light years from Earth. The first Interstellar Radio Message (IRM), the "Arecibo Message", was transmitted in Nov, 1974 from Arecibo Radar Telescope

10.3.2   The Arecibo Message

The Arecibo Observatory (Fig. 10.7) is a radio telescope near the city of Arecibo in Puerto Rico. With a diameter of 305 m it is the largest single dish telescope constructed inside the

Figure 10.7: The Arecibo Radio Telescope, at Arecibo, Puerto Rico. It is the largest dish antenna in the world (305 m). The dish, built into a bowl in the landscape, focuses radio waves from the sky on the feed antenna suspended above it on cables. Since the dish itself can't move, the telescope is steered to point at different regions of the sky by moving the feed antenna (dome) along the curving metal track. Source: wikimedia.

depression left by a karst sinkhole. The telescope has three radar transmitters, with effective isotropic radiated powers of 20 TW at 2380 MHz, 2.5 TW (pulse peak) at 430 MHz, and 300 MW at 47 MHz.

The broadcast was aimed at the globular cluster M13 which is at a distance of 25 000 Ly, so the signal will arrive there in about 25 000 years. The message consisted of 1679 binary digits, approximately 210 bytes, transmitted at a frequency of 2380 MHz and modulated by shifting the frequency by 10 Hz, with a power of 1000 kW. The message consists of seven parts (Fig. 10.8):

- The numbers one (1) through ten (10).

- The atomic numbers of the elements hydrogen, carbon, nitrogen, oxygen, and phosphorus, which make up deoxyribonucleic acid (DNA).

- The formulas for the sugars and bases in the nucleotides of DNA.

- The number of nucleotides in DNA, and a graphic of the double helix structure of DNA.

- A graphic figure of a human, the dimension (physical height) of an average man, and the human population of Earth.

- A graphic of the Solar System.

- A graphic of the Arecibo radio telescope and the dimension (the physical diameter) of the transmitting antenna dish.

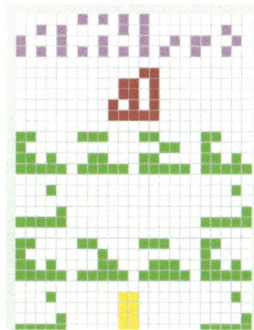

Figure 10.8: The Arecibo message. At the top there are the numbers 1 to 10. Source: wikimedia.

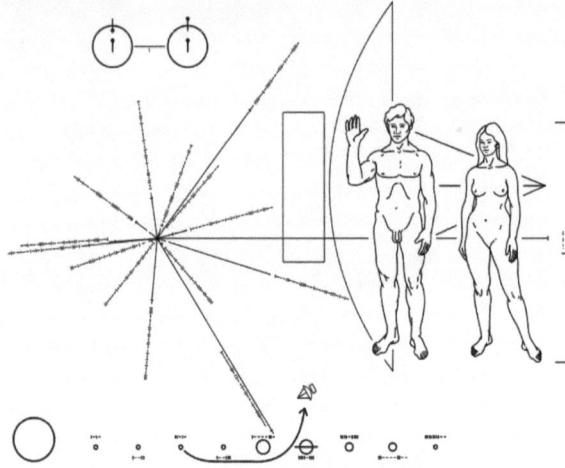

Figure 10.9: The plaque on board the pioneer satellites. Source: wikimedia.

Messages on board of Satellites

The Pioneer plaques (Fig. 10.9) are a pair of gold-anodized aluminium plaques which were placed on board the 1972 Pioneer 10 and 1973 Pioneer 11 spacecraft. The plaques show the nude figures of a human male and female along with several symbols that are designed to provide information about the origin of the spacecraft.

The Pioneer spacecraft were the first human-built objects to leave the Solar System.

The Voyager Golden Records are phonograph records included aboard both Voyager spacecraft, which were launched in 1977. They contain sounds and images selected to portray the diversity of life and culture on Earth, and are intended for any intelligent extraterrestrial life form, or for future humans, who may find them. The Voyager spacecraft is not heading towards any particular star, but Voyager 1 will be within 1.6 light years of the star AC+79 3888 in the Ophiuchus constellation in about 40,000 years. Voyager 1 was launched in 1977, passed the orbit of Pluto in 1990, and then left the Solar System (in the sense of passing the termination shock) in November 2004. It is now in the Kuiper Belt.

10.3.3 Communication with Extraterrestrial Intelligence

There have been seven interstellar radio messages (IRM):

1. The Morse Message (1962). This Soviet experiment was directed to planet Venus and contained the words Mir (Russian Peace), Lenin and SSR.

2. Arecibo Message (1974), one transmission to Messier 13; already described above.

3. Cosmic Call 1 (1999), four transmissions to nearby Sun-like stars The messages were send to the stars liste in Table 10.2.

4. Teen Age Message (2001), six transmissions to the stars listed in Table 10.3

5. Cosmic Call 2 (2003), five transmissions

Table 10.2: Cosmic Call messages.

Name	Designation	Constellation	Sent time	Arrival time	
16 Cyg A	HD 186408	Cygnus	May 24, 1999	Nov 2069	Call 1
15 Sge	HD 190406	Sagitta	June 30, 1999	Feb 2057	Call 1
	HD 178428	Sagitta	June 30, 1999	Oct 2067	Call 1
Gl 777	HD 190360	Cygnus	July 1, 1999	April 2051	Call 1
	Hip 4872	Cassiopeia	July 6, 2003	Apr 2036	Call 2
	HD 245409	Orion	July 6, 2003	Aug 2040	Call 2
55 Cnc	HD 75732	Cancer	July 6, 2003	May 2044	Call 2
	HD 10307	Andromeda	July 6, 2003	Sept 2044	Call 2
47 UMa	HD 95128	Ursa Major	July 6, 2003	May 2049	Call 2

Table 10.3: Teen Age Messages.

Name	HD designation	Constellation	Date sent	Arrival date
	197076	Delphinus	Aug 29, 2001	Feb 2070
47 UMa	95128	Ursa Major	Sept 3, 2001	July 2047
37 Gem	50692	Gemini	Sept 3, 2001	Dec 2057
	126053	Virgo	Sept 3, 2001	Jan 2059
	76151	Hydra	Sept 4, 2001	May 2057
	193664	Draco	Sept 4, 2001	Janu 2059

6. A Message From Earth (2008), one transmission to Gliese 581; sent using the RT-70 radar telescope of Ukraine's National Space Agency. The signal will reach the planet Gliese 581c in early 2029.

7. Across the Universe (2008), a signal was sent to Polaris which is at a distance of 433 Ly.

8. Hello From Earth (HFE, 2009). The message was directed at the nearby red dwarf star Gliese 581 which is at a distance of 20.3 Ly in the constellation of Libra.

10.3.4 Exolinguistics

If there are highly developed alien species how could we communicate with each other? Exolinguistics or xenolinguistics describes a possible language originating from a hypothetical alien species. To understand communication we have to remember that besides spoken words, also non verbal communication plays an important role, even for humans. Messages can be communicated through gestures and touch, by body language or posture, by facial expression and eye contact. Life on Earth employs a variety of non-verbal methods of communication, and these might provide clues to hypothetical alien language.

Amongst other creatures, there are some which use other forms of communication, such as cuttlefish and chameleons, which can alter their body color in complex ways as a method of communication, and ants and honey bees, which use pheromones to communicate complex messages to other members of their hives.

Let us consider ants for example. Ants evolved from wasp-like ancestors in the mid-Cretaceous period between 110 and 130 million years ago. Ants form colonies consisting

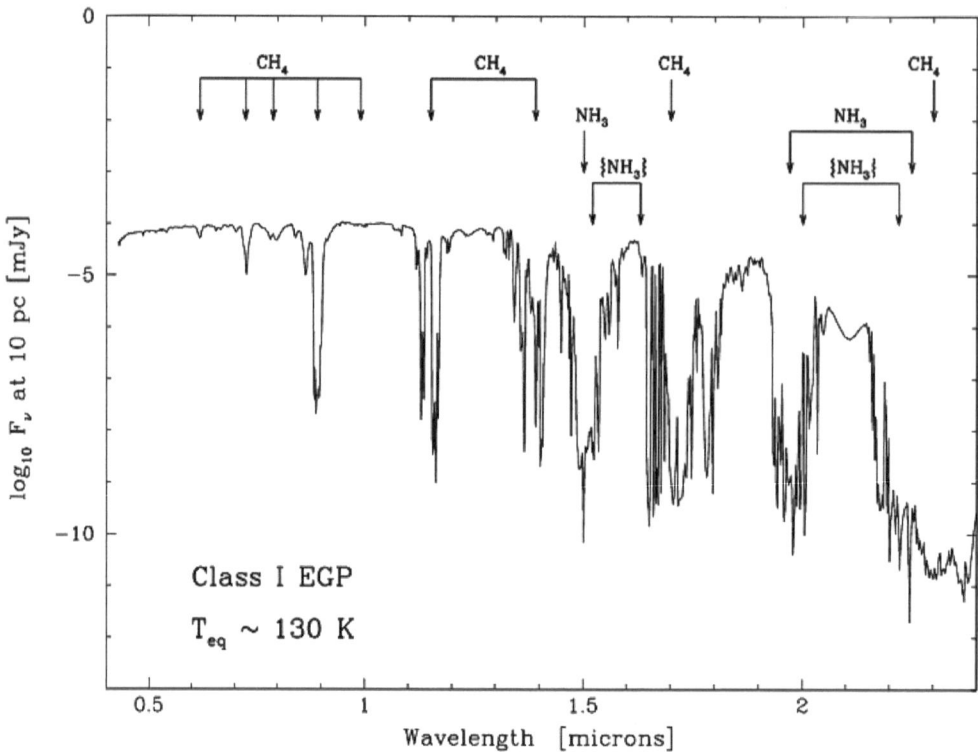

Figure 10.10: Class 1 type spectra of exoplanetary atmospheres. Class 1 denotes cool atmospheres. Source: wikimedia.

of up to millions of individuals. Nearly all ant colonies also have some fertile males called "drones" and one or more fertile females called "queens". Ants communicate with each other using pheromones. Like other insects, ants perceive smells with their long, thin and mobile antennae. When some food is found, they mark a trail, if there is no longer any food, no new marks are deposited and the pheronome scent slowly dissipates. How to find new paths if, for example an established path to a food source is blocked by an obstacle? Then foragers leave the path to explore new routes. If an ant is successful, it leaves a new trail marking the shortest route on its return. Successful trails are followed by more ants, reinforcing better routes and gradually finding the best path.

It is possible that some extraterrestrial species may have no spoken language.

The Dutch mathematician H. Freudenthal of Utrecht University in The Netherlands developed the Lingua Cosmica, Lincos.

10.4   How to Detect Civilizations

We have seen tha direct detection and communciation with extraterrestrial civilizations might be very difficult mainly due to the large distances. We shortly review other possibilities to detect habitated exoplanets.

10.4.1 Biomarkers

Definition: A biomarker is an indicator for a biological state. In biology, a biomarker is a molecule that allows the detection and isolation of a particular cell type. In astrobiology, a biomarker can be any kind of molecule indicating the existence, past or present, of living organisms. In the fields of geology and astrobiology, biomarkers, *versus* geomarkers, are also known as biosignatures.

Typical signatures are:

- complex physical and chemical structures

- utilization of free energy

- production of biomass and wastes.

If there is life on a planet, it will change the atmospheric chemistry: on Earth the typical biosignatures in its atmosphere are oxygen and methane. On Earth methane is created by methanogenesis:

$$CO_2 + 8H + +8e^- \rightarrow CH_4 + 2H_2O \tag{10.2}$$

The presence of atmospheric methane has a role in the scientific search for extra-terrestrial life. The argument being that methane in the atmosphere will eventually dissipate, unless something is replenished in it.

There exist several observation bands for the biomarkers:

- O_3 at 9.6μm

- H_2O at $< 8\,\mu$m

- CO_2 at 15 μ m

- CH_4 at 7.65 μm

- NH_3 at 10.5 μm.

M stars as targets for Terrestrial Exoplanet searches and biosignature detection were discussed in [92]. Oxygen, O_2 plays an important role. The reduction of oxygen provides the largest free energy release per electron transfer, except for the reduction of fluorine and chlorine. However, the bonding of O_2 ensures that it is sufficiently stable to accumulate in a planetary atmosphere, whereas the more weakly bonded halogen gases are far too reactive ever to achieve significant abundance. Consequently, an atmosphere rich in O_2 provides the largest feasible energy source [18]. O_2, O_3 and CH_4 can be detected in the IR (7–25μm. (see also [31]).

10.4.2 Exoplanet Atmospheres

We have already discussed how spectra from exoplanet atmospheres can be obtained. The best chances are observations in the IR since there the intensity contrast between the central star and the planet is lower.

Therefore from observations in these wavelengths different planetary atmospheres are defined depending on the atmospheric temperatures.

- class I (Fig. 10.10) is given for an atmosphere at 130 K,

- class II (T=200 K), see Fig. 10.11,

- class III (T=500 K), see Fig. 10.11,

- class IV (T=1000K), see Fig. 10.12,

- and class V (T=1500 K) in Fig. 10.12.

Activities

Summarize how to detect biologic activity on an exoplanet:

- habitable zone?

- host star?

- planetary parameters?

- location in the Galaxy?

- ....

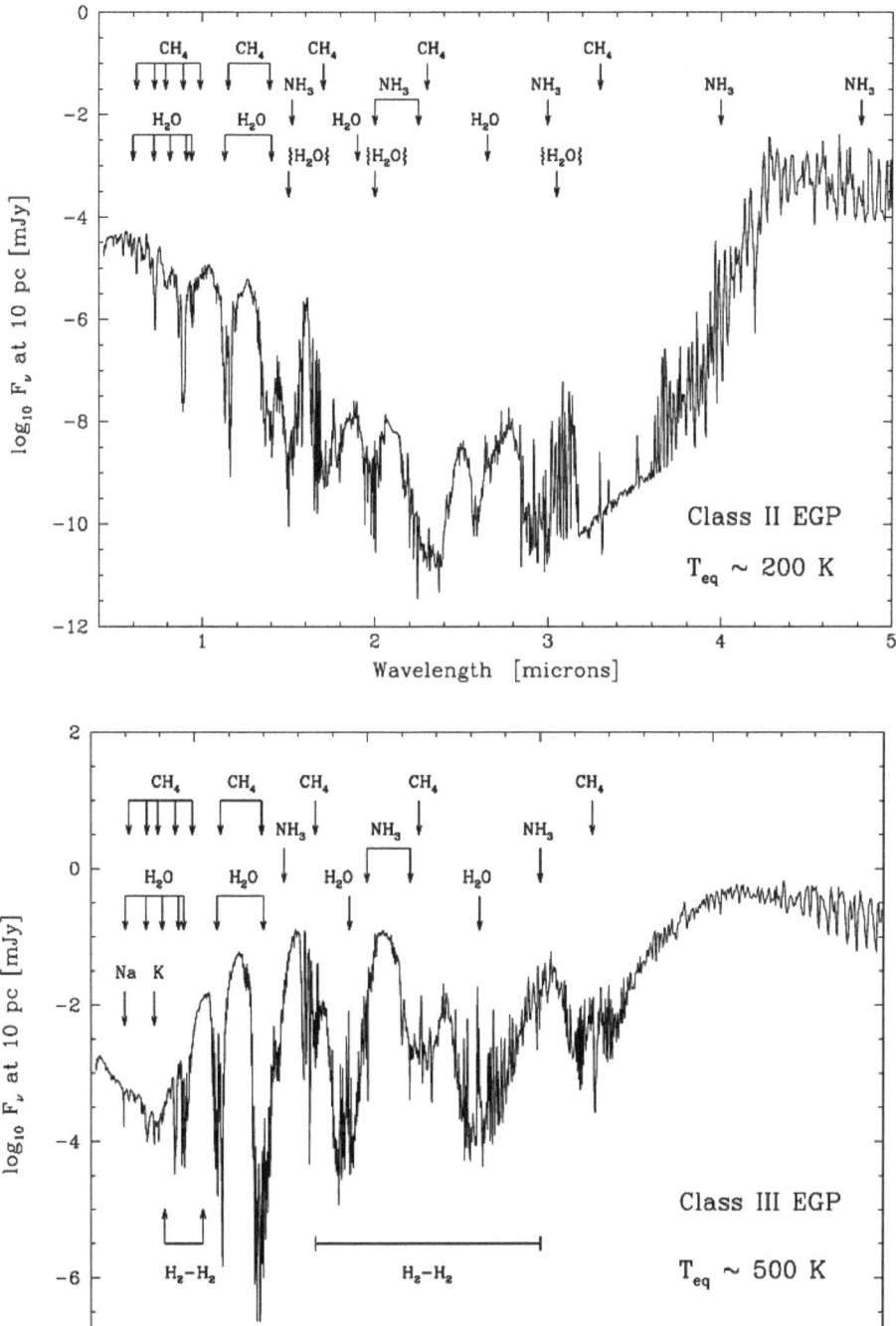

Figure 10.11: Class 2 and class 3 type spectra of exoplanetary atmospheres. Class 2 and 3 denote atmospheres with moderate temperatures (less than 500 K). Source: wikimedia.

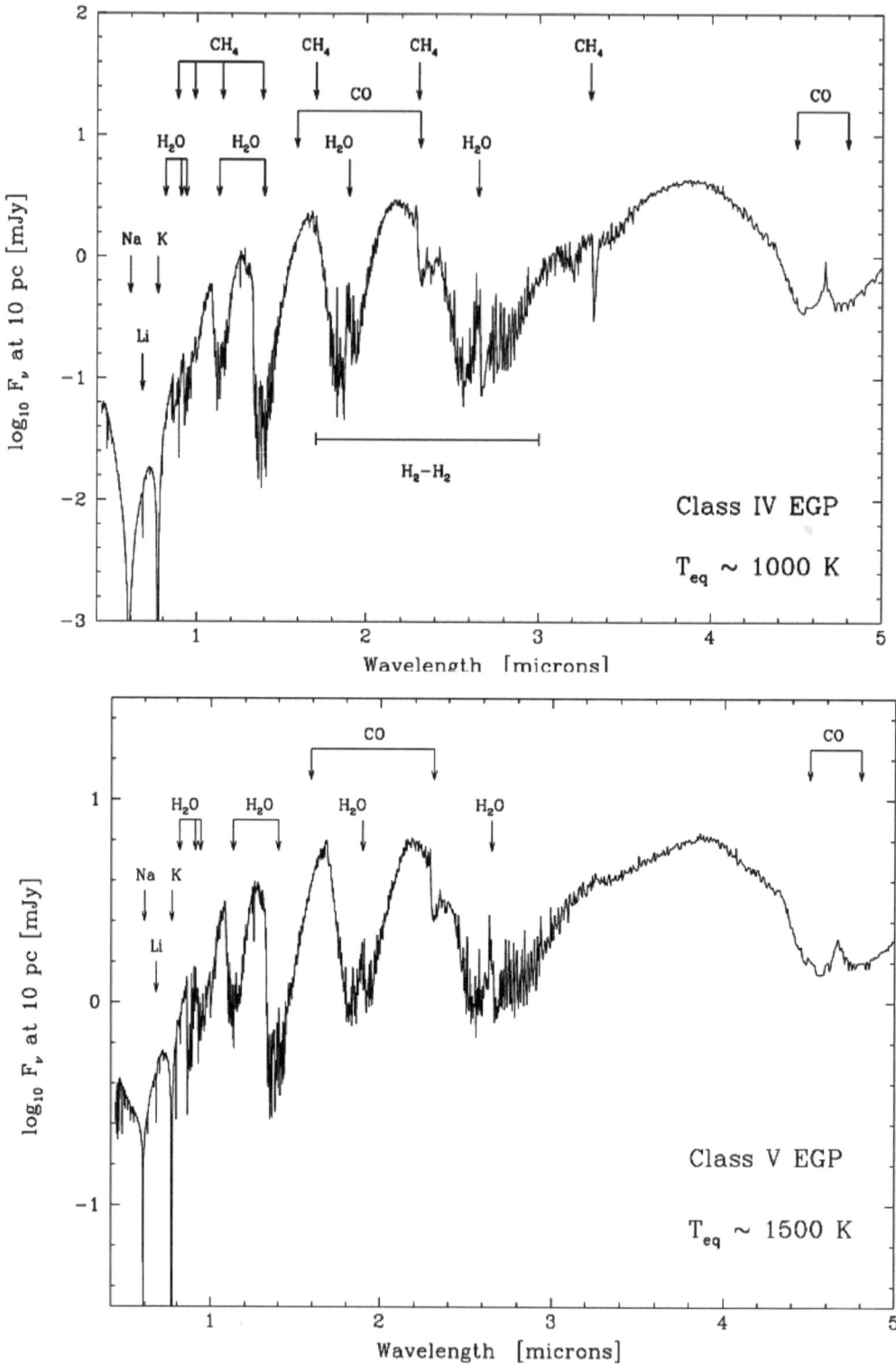

Figure 10.12: Class 4 and class 5 type spectra of exoplanetary atmospheres. Class 4 and 5 denote high temperature atmospheres ($T > 1000\,\mathrm{K}$). Source: wikimedia.

Bibliography

[1] G. Anglada-Escudé, P. Arriagada, S. S. Vogt, E. J. Rivera, R. P. Butler, J. D. Crane, S. A. Shectman, I. B. Thompson, D. Minniti, N. Haghighipour, B. D. Carter, C. G. Tinney, R. A. Wittenmyer, J. A. Bailey, S. J. O'Toole, H. R. A. Jones, and J. S. Jenkins. A planetary system around the nearby M dwarf GJ 667C with at least one super-Earth in its habitable zone. *ArXiv e-prints*, February 2012.

[2] J. A. Baross. Growth of 'black smoker' bacteria at temperatures of at least 250 ^{0}C. *Nature*, 303:423–426, June 1983.

[3] J.-A. Barrat and C. Bollinger. Geochemistry of the Martian meteorite ALH84001, revisited. *Meteoritics and Planetary Science*, 45:495–512, April 2010.

[4] A. T. Basilevsky, M. A. Ivanov, J. W. Head, M. Aittola, and J. Raitala. Landing on Venus: Past and future. *Planetary and Space Science*, 55:2097–2112, November 2007.

[5] N. M. Batalha, W. J. Borucki, S. T. Bryson, L. A. Buchhave, D. A. Caldwell, J. Christensen-Dalsgaard, D. Ciardi, E. W. Dunham, F. Fressin, T. N. Gautier, III, R. L. Gilliland, M. R. Haas, S. B. Howell, J. M. Jenkins, H. Kjeldsen, D. G. Koch, D. W. Latham, J. J. Lissauer, G. W. Marcy, J. F. Rowe, D. D. Sasselov, S. Seager, J. H. Steffen, G. Torres, G. S. Basri, T. M. Brown, D. Charbonneau, J. Christiansen, B. Clarke, W. D. Cochran, A. Dupree, D. C. Fabrycky, D. Fischer, E. B. Ford, J. Fortney, F. R. Girouard, M. J. Holman, J. Johnson, H. Isaacson, T. C. Klaus, P. Machalek, A. V. Moorehead, R. C. Morehead, D. Ragozzine, P. Tenenbaum, J. Twicken, S. Quinn, J. VanCleve, L. M. Walkowicz, W. F. Welsh, E. Devore, and A. Gould. Kepler's First Rocky Planet: Kepler-10b. *Astrophysical Journal*, 729:27, March 2011.

[6] L. Battison. Niche Habitats for Extra-Terrestrial Life: The Potential for Astrobiology on the Moons of Saturn and Jupiter. *Journal of Cosmology*, 13:3890, March 2011.

[7] G. F. Benedict, B. E. McArthur, T. Forveille, X. Delfosse, E. Nelan, R. P. Butler, W. Spiesman, G. Marcy, B. Goldman, C. Perrier, W. H. Jefferys, and M. Mayor. A Mass for the Extrasolar Planet Gliese 876b Determined from Hubble Space Telescope Fine Guidance Sensor 3 Astrometry and High-Precision Radial Velocities. *Astrophysical Journal Letters*, 581:L115–L118, December 2002.

[8] G. F. Benedict, B. E. McArthur, G. Gatewood, E. Nelan, W. D. Cochran, A. Hatzes, M. Endl, R. Wittenmyer, S. L. Baliunas, G. A. H. Walker, S. Yang, M. Krster, S. Els, and D. B. Paulson. The Extrasolar Planet ϵ Eridani b: Orbit and Mass. *Astronomical Journal*, 132:2206–2218, November 2006.

[9] H. Beust, D. Bonneau, D. Mourard, S. Lafrasse, G. Mella, G. Duvert, and A. Chelli. On the use of the Virtual Observatory to select calibrators for phase-referenced astrometry of exoplanet-host stars. *Monthly Notices*, 414:108–115, June 2011.

[10] K. Biemann, J. Oro, P. Toulmin, III, L. E. Orgel, A. O. Nier, D. M. Anderson, D. Flory, A. V. Diaz, D. R. Rushneck, and P. G. Simmonds. The search for organic substances and inorganic volatile compounds in the surface of Mars. *Journal of Geophys. Research*, 82:4641–4658, September 1977.

[11] A. C. Boley, M. J. Payne, S. Corder, W. R. F. Dent, E. B. Ford, and M. Shabram. Constraining the Planetary System of Fomalhaut Using High-resolution ALMA Observations. *Astrophysical Journal Letters*, 750:L21, May 2012.

[12] V. Bromm and N. Yoshida. The First Galaxies. *Annual Review of Astronomy and Astrophysics*, 49:373–407, September 2011.

[13] M. J. Burchell. W(h)ither the Drake equation? *International Journal of Astrobiology*, 5:243–250, December 2006.

[14] K. L. Cahoy, M. S. Marley, and J. J. Fortney. Exoplanet Albedo Spectra and Colors as a Function of Planet Phase, Separation, and Metallicity. *Astrophysical Journal*, 724:189–214, November 2010.

[15] R. W. Carlson. A Tenuous Carbon Dioxide Atmosphere on Jupiter's Moon Callisto. *Science*, 283:820, February 1999.

[16] J. S. Carr and J. R. Najita. Organic Molecules and Water in the Inner Disks of T Tauri Stars. *Astrophysical Journal*, 733:102, June 2011.

[17] T. A. Cassidy. *Europa's tenuous atmosphere*. PhD thesis, University of Virginia, 2008.

[18] D. C. Catling, C. R. Glein, K. J. Zahnle, and C. P. McKay. Why O_2 Is Required by Complex Life on Habitable Planets and the Concept of Planetary "Oxygenation Time". *Astrobiology*, 5:415–438, June 2005.

[19] R. Cavicchioli. Extremophiles and the Search for Extraterrestrial Life. *Astrobiology*, 2:281–292, August 2002.

[20] E. Chassefière, R. Wieler, B. Marty, and F. Leblanc. The evolution of Venus: Present state of knowledge and future exploration. *Planetary and Space Science*, 63:15–23, April 2012.

[21] G. Chauvin, A.-M. Lagrange, B. Zuckerman, C. Dumas, D. Mouillet, I. Song, J.-L. Beuzit, P. Lowrance, and M. S. Bessell. A companion to AB Pic at the planet/brown dwarf boundary. *Astronomy and Astrophysics*, 438:L29–L32, August 2005.

[22] C. C. Christopher, Y. L. Yung, M. C. Liang, H. Hartman, C. J. Hansen, G. Tinetti, V. Meadows, and J. L. Kirschvink. Enceladus: Cassini Observations and Implications for the Search for Life. *AGU Fall Meeting Abstracts*, page B176, December 2006.

[23] B. Ciardi and A. Ferrara. The First Cosmic Structures and Their Effects. *r*, 116:625–705, February 2005.

[24] M. M. Cirkovic. Fermi's Paradox - The Last Challenge For Copernicanism? *Serbian Astronomical Journal*, 178:1–20, June 2009.

[25] C. S. Cockell. Life on Venus. *Planetary and Space Science*, 47:1487–1501, December 1999.

[26] C. S. Cockell. Astrobiology–What Can We Do on the Moon? *Earth Moon and Planets*, 107:3–10, December 2010.

[27] V. Cottini, N. I. Ignatiev, G. Piccioni, P. Drossart, D. Grassi, and W. J. Markiewicz. Water vapor near the cloud tops of Venus from Venus Express/VIRTIS dayside data. *Icarus*, 217:561–569, February 2012.

[28] M. Cuntz, W. von Bloh, K.-P. Schrder, C. Bounama, and S. Franck. Habitability of super-Earth planets around main-sequence stars including red giant branch evolution: models based on the integrated system approach. *International Journal of Astrobiology*, 11:15–23, January 2012.

[29] E. J. W. de Mooij, M. Brogi, R. J. de Kok, J. Koppenhoefer, S. V. Nefs, I. A. G. Snellen, J. Greiner, J. Hanse, R. C. Heinsbroek, C. H. Lee, and P. P. van der Werf. Optical to near-infrared transit observations of super-Earth GJ 1214b: water-world or mini-Neptune? *Astronomy and Astrophysics*, 538:A46, February 2012.

[30] A. H. Delsemme. Cometary origin of carbon, nitrogen and water on the Earth. *Origins of Life and Evolution of the Biosphere*, 21:279–298, September 1991.

[31] D. J. Des Marais, M. O. Harwit, K. W. Jucks, J. F. Kasting, D. N. C. Lin, J. I. Lunine, J. Schneider, S. Seager, W. A. Traub, and N. J. Woolf. Remote Sensing of Planetary Properties and Biosignatures on Extrasolar Terrestrial Planets. *Astrobiology*, 2:153–181, June 2002.

[32] A. G. Fairén, J. M. Dohm, V. R. Baker, S. D. Thompson, W. C. Mahaney, K. E. Herkenhoff, J. A. P. Rodríguez, A. F. Davila, D. Schulze-Makuch, M. R. El Maarry, E. R. Uceda, R. Amils, H. Miyamoto, K. J. Kim, R. C. Anderson, and C. P. McKay. Meteorites at Meridiani Planum provide evidence for significant amounts of surface and near-surface water on early Mars. *Meteoritics and Planetary Science*, 46:1832–1841, December 2011.

[33] A. D. Fortes. Titan's internal structure and the evolutionary consequences. *Planetary and Space Science*, 60:10–17, January 2012.

[34] F. Fressin, G. Torres, J. F. Rowe, D. Charbonneau, L. A. Rogers, S. Ballard, N. M. Batalha, W. J. Borucki, S. T. Bryson, L. A. Buchhave, D. R. Ciardi, J.-M. Désert, C. D. Dressing, D. C. Fabrycky, E. B. Ford, T. N. Gautier, III, C. E. Henze, M. J. Holman, A. Howard, S. B. Howell, J. M. Jenkins, D. G. Koch, D. W. Latham, J. J. Lissauer, G. W. Marcy, S. N. Quinn, D. Ragozzine, D. D. Sasselov, S. Seager, T. Barclay, F. Mullally, S. E. Seader, M. Still, J. D. Twicken, S. E. Thompson, and K. Uddin. Two Earth-sized planets orbiting Kepler-20. *Nature*, 482:195–198, February 2012.

[35] M. S. Giampapa, L. Golub, R. Rosner, G. S. Vaiana, J. L. Linsky, and S. P. Worden. A heating mechanism for the chromospheres of M dwarf stars. *SAO Special Report*, 392:A73, February 1982.

[36] J. P. Greenwood, S. Itoh, N. Sakamoto, P. Warren, L. Taylor, and H. Yurimoto. Hydrogen isotope ratios in lunar rocks indicate delivery of cometary water to the Moon. *Nature Geoscience*, 4:79–82, February 2011.

[37] R. M. Haberle, M. A. Kahre, J. L. Hollingsworth, J. Schaeffer, F. Montmessin, and R. J. Phillips. A Cloud Greenhouse Effect on Mars: Significant Climate Change in the Recent Past? In *Lunar and Planetary Institute Science Conference Abstracts*, volume 43 of *Lunar and Planetary Institute Science Conference Abstracts*, page 1665, March 2012.

[38] J. C. Hall and G. W. Lockwood. Evidence of a Pronounced Activity Cycle in the Solar Twin 18 Scorpii. *Astrophysical Journal Letters*, 545:L43–L45, December 2000.

[39] K. P. Hand. Is there life on Europa? *Nature*, 457:384–385, January 2009.

[40] K. P. Hand, C. F. Chyba, J. C. Priscu, R. W. Carlson, and K. H. Nealson. *Astrobiology and the Potential for Life on Europa*, page 589. 2009.

[41] A. Hanslmeier. *Habitability and Cosmic Catastrophes*. 2009.

[42] A. Hanslmeier, editor. *Water in the Universe*, volume 368 of *Astrophysics and Space Science Library*, 2011.

[43] H. G. M. Hill, I. Gilmour, V. K. Pearson, and J. A. Nuth. Did Organic Compounds in the Tagish Lake Meteorite Form Via Catalytic Processes in the Solar Nebula and Within Parent Bodies? *Meteoritics and Planetary Science Supplement*, 38:5038, July 2003.

[44] C. M. Huitson, D. K. Sing, A. Vidal-Madjar, G. E. Ballester, A. Lecavelier des Etangs, J.-M. Désert, and F. Pont. Temperature-Pressure Profile of the hot Jupiter HD 189733b from HST Sodium Observations: Detection of Upper Atmospheric Heating. *ArXiv e-prints*, February 2012.

[45] R. Jaumann, R. L. Kirk, R. D. Lorenz, R. M. C. Lopes, E. Stofan, E. P. Turtle, H. U. Keller, C. A. Wood, C. Sotin, L. A. Soderblom, and M. G. Tomasko. *Geology and Surface Processes on Titan*, page 75. 2010.

[46] R. E. Johnson, T. A. Cassidy, F. Leblanc, A. Hendrix, M. Marconi, F. Cipriani, and M. Wong. Atmospheres of Europa and Ganymede. In *European Planetary Science Congress 2010*, page 246, September 2010.

[47] J. F. Kasting, D. P. Whitmire, and R. T. Reynolds. Habitable Zones around Main Sequence Stars. *Icarus*, 101:108–128, January 1993.

[48] D. S. Kelley, J. A. Baross, and J. R. Delaney. Volcanoes, Fluids, and Life at Mid-Ocean Ridge Spreading Centers. *Annual Review of Earth and Planetary Sciences*, 30:385–491, 2002.

[49] M. G. Kendrick and T. A. Kral. Survival of Methanogens During Desiccation: Implications for Life on Mars. *Astrobiology*, 6:546–551, August 2006.

[50] K. K. Khurana, X. Jia, M. G. Kivelson, F. Nimmo, G. Schubert, and C. T. Russell. Evidence of a global subsurface magma ocean in Io from Galileo's magnetometer observations. In *EPSC-DPS Joint Meeting 2011*, page 751, October 2011.

[51] K. K. Khurana, X. Jia, M. G. Kivelson, and R. J. Walker. Magnetic induction Signals at Ganymede: implications for a subsurface ocean. In *European Planetary Science Congress 2009*, page 335, September 2009.

[52] K. K. Khurana, M. G. Kivelson, D. J. Stevenson, G. Schubert, C. T. Russell, R. J. Walker, and C. Polanskey. Induced magnetic fields as evidence for subsurface oceans in Europa and Callisto. *Nature*, 395:777–780, October 1998.

[53] S. Kounaves. Life on Mars may be hidden like Earth's extremophiles. *Nature*, 449:281, September 2007.

[54] T. A. Kral, T. S. Altheide, A. E. Lueders, T. H. Goodhart, B. T. Virden, W. Birch, K. L. Howe, and P. Gavin. Methanogens: A Model for Life on Mars. *LPI Contributions*, 1538:5084, April 2010.

[55] L. V. Ksanfomaliti. Planetary Systems around Stars of Late Spectral Types: A Limitation for Habitable Zones. *Solar System Research*, 32:413, 1998.

[56] G. P. Kuiper. Infrared observations of planets and satellites. *Astronomical Journal*, 62:245, November 1957.

[57] A. K. Lal. Origin of Life. *Astrophysics and Space Science*, 317:267–278, October 2008.

[58] H. Lammer, J. H. Bredehft, A. Coustenis, M. L. Khodachenko, L. Kaltenegger, O. Grasset, D. Prieur, F. Raulin, P. Ehrenfreund, M. Yamauchi, J.-E. Wahlund, J.-M. Griemeier, G. Stangl, C. S. Cockell, Y. N. Kulikov, J. L. Grenfell, and H. Rauer. What makes a planet habitable? *The Astronomy and Astrophysics Review*, 17:181–249, June 2009.

[59] B. Langlais, E. Thébault, E. Ostanciaux, and N. Mangold. A Late Martian Dynamo Cessation Time 3.77 Gy Ago. In *Lunar and Planetary Institute Science Conference Abstracts*, volume 43 of *Lunar and Planetary Inst. Technical Report*, page 1231, March 2012.

[60] J.-P. Lebreton, A. Coustenis, J. Lunine, F. Raulin, T. Owen, and D. Strobel. Results from the Huygens probe on Titan. *The Astronomy and Astrophysics Review*, 17:149–179, June 2009.

[61] E. Lellouch, M. A. McGrath, and K. L. Jessup. *Io's atmosphere*, page 231. Springer Praxis Books / Geophysical Sciences, 2007.

[62] M.-C. Liang, B. F. Lane, R. T. Pappalardo, M. Allen, and Y. L. Yung. Atmosphere of Callisto. *Journal of Geophysical Research (Planets)*, 110:2003, February 2005.

[63] G. F. Lindal, G. E. Wood, G. S. Levy, J. D. Anderson, D. N. Sweetnam, H. B. Hotz, B. J. Buckles, D. P. Holmes, P. E. Doms, V. R. Eshleman, G. L. Tyler, and T. A. Croft. The atmosphere of Jupiter - an analysis of the Voyager radio occultation measurements. *Journal of Geophys. Research*, 86:8721–8727, September 1981.

[64] C. Lineweaver. The Galactic Habitable Zone and the Age Distribution of Complex Life in the Milky Way. In *American Astronomical Society Meeting Abstracts #210*, volume 38 of *Bulletin of the American Astronomical Society*, page 173, May 2007.

[65] C. H. Lineweaver, Y. Fenner, and B. K. Gibson. The Galactic Habitable Zone and the Age Distribution of Complex Life in the Milky Way. *Science*, 303:59–62, January 2004.

[66] K. Lodders and B. Fegley. *The planetary scientist's companion / Katharina Lodders, Bruce Fegley.* The planetary scientist's companion / Katharina Lodders, Bruce Fegley. New York : Oxford University Press, 1998. QB601 .L84 1998, 1998.

[67] P. López-García. *Extremophiles*, page 657. Springer-Verlag, 2005.

[68] J. Lunine, E. Stofan, C. Elachi, R. Lorenz, B. Stiles, K. L. Mitchell, S. Ostro, L. Soderblom, C. Wood, H. Zebker, S. Wall, M. Janssen, R. Kirk, R. Lopes, F. Paganelli, J. Radebaugh, L. Wye, P. Callahan, Y. Anderson, M. Allison, R. Boehmer, P. Encrenaz, E. Flamini, G. Franceschetti, Y. Gim, G. Hamilton, S. Hensely, W. T. Johnson, K. Kelleher, D. Muhleman, P. Paillou, G. Picardi, F. Posa, L. Roth, R. Seu, S. Shaffer, S. Vetrella, R. West, and R. Orosei. The Lakes of Titan. *AGU Fall Meeting Abstracts*, page A5, December 2006.

[69] R. L. Mancinelli. Peroxides and the survivability of microorganisms on the surface of Mars. *Advances in Space Research*, 9:191–195, 1989.

[70] A. M. Mandell, J. Bast, E. F. van Dishoeck, G. A. Blake, C. Salyk, M. J. Mumma, and G. Villanueva. First Detection of Near-infrared Line Emission from Organics in Young Circumstellar Disks. *Astrophysical Journal*, 747:92, March 2012.

[71] C. Marois, B. Zuckerman, Q. M. Konopacky, B. Macintosh, and T. Barman. Images of a fourth planet orbiting HR 8799. *Nature*, 468:1080–1083, December 2010.

[72] M. Mayor and D. Queloz. A Jupiter-mass companion to a solar-type star. *Nature*, 378:355–359, November 1995.

[73] T. B. McCord, R. Carlson, W. Smythe, G. Hansen, R. Clark, C. Hibbitts, F. Fanale, J. Granahan, M. Segura, D. Matson, T. Johnson, and P. Martin. Organics and other molecules in the surfaces of Callisto and Ganymede. *Science*, 278:271–275, October 1997.

[74] J. C. Morales, E. Herrero, I. Ribas, and C. Jordi. Earth-like planets around M-type stars. In M. R. Zapatero Osorio, J. Gorgas, J. Maíz Apellániz, J. R. Pardo, and A. Gil de Paz, editors, *Highlights of Spanish Astrophysics VI*, pages 627–627, November 2011.

[75] H. Morowitz. Life on Venus. *Astrobiology*, 11:931–932, November 2011.

[76] P. Odert, M. Leitzinger, A. Hanslmeier, H. Lammer, M. L. Khodachenko, and I. Ribas. M-Type Stars as Hosts for Habitable Planets. In V. Coudé Du Foresto, D. M. Gelino, and I. Ribas, editors, *Pathways Towards Habitable Planets*, volume 430 of *Astronomical Society of the Pacific Conference Series*, page 515, October 2010.

[77] J. Pap, M. Anklin, C. Fröhlich, C. Wehrli, F. Varadi, and L. Floyd. Variations in total solar and spectral irradiance as measured by the VIRGO experiment on SOHO. *Advances in Space Research*, 24:215–224, 1999.

[78] E. T. Parker, H. J. Cleaves, M. P. Callahan, J. P. Dworkin, D. P. Glavin, A. Lazcano, and J. L. Bada. Prebiotic Synthesis of Methionine and Other Sulfur-Containing Organic Compounds on the Primitive Earth: A Contemporary Reassessment Based on an Unpublished 1958 Stanley Miller Experiment. *Origins of Life and Evolution of the Biosphere*, 41:201–212, June 2011.

[79] E. Pierazzo and C. F. Chyba. Cometary Delivery of Biogenic Elements to Europa. *Icarus*, 157:120–127, May 2002.

[80] P. C. Plait. The Face on Mars. *Sky and Telescope*, 113(2):020000, February 2007.

[81] G. Porto de Mello, R. da Silva, and L. da Silva. A Survey of Solar Twin Stars within 50 Parsecs of the Sun. In G. Lemarchand & K. Meech, editor, *Bioastronomy 99*, volume 213 of *Astronomical Society of the Pacific Conference Series*, page 73, 2000.

[82] O. Prieto-Ballesteros, E. Vorobyova, V. Parro, J. A. Rodriguez Manfredi, and F. Gómez. Strategies for detection of putative life on Europa. *Advances in Space Research*, 48:678–688, August 2011.

[83] Z. Regály, Z. Sándor, C. P. Dullemond, and R. van Boekel. Detectability of giant planets in protoplanetary disks by CO emission lines. *Astronomy and Astrophysics*, 523:A69, November 2010.

[84] E. M. Rivkina, V. A. Sherbakova, K. V. Krivushin, A. A. Abramov, and D. A. Gilichinsky. Permafrost on Earth – Models and Analogues of Martian Habitats and Inhabitants. *LPI Contributions*, 1538:5620, April 2010.

[85] B. E. Robertson, R. S. Ellis, J. S. Dunlop, R. J. McLure, and D. P. Stark. Early star-forming galaxies and the reionization of the Universe. *Nature*, 468:49–55, November 2010.

[86] C. K. Robinson and J. Diruggiero. Damage Avoidance and DNA Repair Mechanisms of Extremophiles to Ionizing Radiation. *LPI Contributions*, 1538:5556, April 2010.

[87] L. A. Rogers, P. Bodenheimer, J. J. Lissauer, and S. Seager. Formation and Structure of Low-density exo-Neptunes. *Astrophysical Journal*, 738:59, September 2011.

[88] L. Rothschild. *Extremophiles: defining the envelope for the search for life in the universe*, page 113. Cambridge University Press, 2007.

[89] L. J. Rothschild and R. L. Mancinelli. Life in extreme environments. *Nature*, 409:1092–1101, February 2001.

[90] C. Sagan. Life on the Surface of Venus? *Nature*, 216:1198–1199, December 1967.

[91] J. S. Saur, P. D. Feldman, D. F. Strobel, K. D. Retherford, L. Roth, M. A. McGrath, J. M. Gerard, D. C. Grodent, and N. Schilling. HST observations of Europa's atmospheric UV emission. *AGU Fall Meeting Abstracts*, December 2009.

[92] J. Scalo, L. Kaltenegger, A. G. Segura, M. Fridlund, I. Ribas, Y. N. Kulikov, J. L. Grenfell, H. Rauer, P. Odert, M. Leitzinger, F. Selsis, M. L. Khodachenko, C. Eiroa, J. Kasting, and H. Lammer. M Stars as Targets for Terrestrial Exoplanet Searches And Biosignature Detection. *Astrobiology*, 7:85–166, February 2007.

[93] A. C. Schuerger, J. E. Moores, C. Clausen, N. G. Barlow, and D. T. Britt. A Proposed UV/CH4 Linked Model for the Global Methane Budget on Mars. In *Lunar and Planetary Institute Science Conference Abstracts*, volume 43 of *Lunar and Planetary Institute Science Conference Abstracts*, page 1911, March 2012.

[94] D. Schulze-Makuch. Io: Is Life Possible Between Fire and Ice? *Journal of Cosmology*, 5:912–919, February 2010.

[95] D. Schulze-Makuch, D. H. Grinspoon, O. Abbas, L. N. Irwin, and M. A. Bullock. A Sulfur-Based Survival Strategy for Putative Phototrophic Life in the Venusian Atmosphere. *Astrobiology*, 4:11–18, March 2004.

[96] J. Seckbach and J. Chela-Flores. Extremophiles and chemotrophs as contributors to astrobiological signatures on Europa: a review of biomarkers of sulfate-reducers and other microorganisms. In *Society of Photo-Optical Instrumentation Engineers (SPIE) Conference Series*, volume 6694 of *Society of Photo-Optical Instrumentation Engineers (SPIE) Conference Series*, October 2007.

[97] F. Selsis, J. F. Kasting, B. Levrard, J. Paillet, I. Ribas, and X. Delfosse. Habitable planets around the star Gliese 581? *Astronomy and Astrophysics*, 476:1373–1387, December 2007.

[98] G. Southam, L. J. Rothschild, and F. Westall. The Geology and Habitability of Terrestrial Planets: Fundamental Requirements for Life. *r*, 129:7–34, March 2007.

[99] D. S. Spiegel, A. Burrows, L. Ibgui, I. Hubeny, and J. A. Milsom. Models of Neptune-Mass Exoplanets: Emergent Fluxes and Albedos. *Astrophysical Journal*, 709:149–158, January 2010.

[100] C. R. Stoker and M. A. Bullock. Organic degradation under simulated Martian conditions. *Journal of Geophys. Research*, 102:10881–10888, May 1997.

[101] R. G. Strom, R. J. Terrile, C. Hansen, and H. Masursky. Volcanic eruption plumes on Io. *Nature*, 280:733–736, August 1979.

[102] M. Sundin. The galactic habitable zone in barred galaxies. *International Journal of Astrobiology*, 5:325–326, September 2006.

[103] K. L. Thomas-Keprta, D. A. Bazylinski, J. L. Kirschvink, S. J. Clemett, D. S. McKay, S. J. Wentworth, H. Vali, E. K. Gibson, and C. S. Romanek. Elongated prismatic magnetite crystals in ALH84001 carbonate globules: - Potential Martian magnetofossils. *Geochimica et Cosmochimica Acta*, 64:4049–4081, December 2000.

[104] G. Tinetti, M. Liang, J. Beaulieu, Y. L. Yung, S. Carey, I. Ribas, J. Tennyson, B. Barber, N. Allard, G. Ballester, D. Sing, and F. Selsis. Water Vapour In The Atmosphere Of An Extrasolar Planet. In *AAS/Division for Planetary Sciences Meeting Abstracts #39*, volume 38 of *Bulletin of the American Astronomical Society*, page 467, October 2007.

[105] M. G. Trainer, A. A. Pavlov, H. L. Dewitt, J. L. Jimenez, C. P. McKay, O. B. Toon, and M. A. Tolbert. Inaugural Article: Organic haze on Titan and the early Earth. *Proceedings of the National Academy of Science*, 103:18035–18042, November 2006.

[106] K. L. Von Damm. Seafloor Hydrothermal Activity: Black Smoker Chemistry and Chimneys. *Annual Review of Earth and Planetary Sciences*, 18:173, 1990.

[107] B. Vukotic and M. M. Cirkovic. On the Timescale Forcing in Astrobiology. *Serbian Astronomical Journal*, 175:45–50, December 2007.

[108] G. Wchtershuser. The case for an autotrophic origin. *Origins of Life and Evolution of the Biosphere*, 19:423–424, May 1989.

[109] G. Wchtershuser. The case for the chemoautotrophic origin of life in an iron-sulfur world. *Origins of Life and Evolution of the Biosphere*, 20:173–176, March 1990.

[110] P. S. Wesson. Panspermia, Past and Present: Astrophysical and Biophysical Conditions for the Dissemination of Life in Space. *r*, 156:239–252, October 2010.

[111] A. Wolszczan and M. Kuchner. *Planets Around Pulsars and Other Evolved Stars: The Fates of Planetary Systems*, pages 175–190. 2010.

[112] J. I. Zuluaga and P. A. Cuartas. The role of rotation in the evolution of dynamo-generated magnetic fields in Super Earths. *Icarus*, 217:88–102, January 2012.

Index

www.ingramcontent.com/pod-product-compliance
Lightning Source LLC
Chambersburg PA
CBHW050840220326
41598CB00006B/409